AF360913

Air Pollution and Plant Ecosystems

Air Pollution and Plant Ecosystems

Editors

Evgenios Agathokleous
Elisa Carrari
Pierre Sicard

MDPI • Basel • Beijing • Wuhan • Barcelona • Belgrade • Manchester • Tokyo • Cluj • Tianjin

Editors
Evgenios Agathokleous
Institute of Ecology, School of
Applied Meteorology,
Nanjing University of
Information Science &
Technology
China

Elisa Carrari
Consiglio Nazionale delle
Ricerche, Institute for
Sustainable Plant Protection
Italy

Pierre Sicard
ARGANS, Sophia Antipolis
France

Editorial Office
MDPI
St. Alban-Anlage 66
4052 Basel, Switzerland

This is a reprint of articles from the Special Issue published online in the open access journal *Climate* (ISSN 2225-1154) (available at: https://www.mdpi.com/journal/climate/special_issues/ozone_plant_ecosystems).

For citation purposes, cite each article independently as indicated on the article page online and as indicated below:

LastName, A.A.; LastName, B.B.; LastName, C.C. Article Title. *Journal Name* **Year**, *Article Number*, Page Range.

ISBN 978-3-03943-284-4 (Hbk)
ISBN 978-3-03943-285-1 (PDF)

Contents

About the Editors

Evgenios Agathokleous graduated from Agricultural University of Athens (AUA), Greece, in February 2013 with a Diploma in Agriculture (equivalent to the Anglo-Saxon MSc in Agricultural Science). As a scholar of the Government of Japan, he continued his studies at Hokkaido University in Sapporo, Japan. From April to September 2014, he was a Research Student at the Research Faculty of Agriculture. From October 2014, he was a PhD student at the same School. He graduated from the Special Postgraduate Program in Biosphere Sustainability Science with a PhD in Environmental Resources in March 2017. During the period April 2017–March 2019, he was an International Research Fellow of the Japan Society for the Promotion of Science, hosted by Forestry and Forest Products Research Institute. At the same time, he was a Researcher at the Research Faculty of Agriculture of Hokkaido University. Since April 2019, he has served as Full Professor at the School of Applied Meteorology at Nanjing University of Information Science and Technology in Nanjing, China. His research concerns the dose–response relationship and its mechanisms. He has so far published around 90 articles in prestigious international scientific journals, including about 50 as first author. Some of his papers are published in journals with an impact factor in the range 12–20 (e.g., Nano Today, Science Advances, Trends in Plant Science, Trends in Pharmacology), and including invited papers in several journals. He has also authored 10 book chapters, of which 9 are international. His resume includes 34 oral and 49 poster presentations; almost all delivered at international conferences. He has been honored with the Outstanding New Investigator 2018 award of the International Dose–Response Society for his research on hormesis. He has been involved in large-scale research funding, with a significant portion of it ensured either as a fellow or as a Principal Investigator of projects. He has reviewed 390 papers for 62 SCI journals. He is Associate Editor-in-Chief of the Journal of Forestry Research (Springer) and Editor of Science of the Total Environment (Elsevier), Plant Stress (Elsevier), Climate (MDPI), Frontiers in Forests and Global Change (Frontiers), and Sci (MDPI).

Elisa Carrari started her academic studies at the Agriculture faculty of the University of Florence (Italy) where she graduated in 2012 in Science and Technology of Forest Systems. In 2016, she received her PhD in Plant, Microbiology, and Genetic Science and Technology, with the certification of Doctor Europaeus from the same university in co-supervision with the University of Ghent (Belgium). She was declared expert in "Biodiversity of Forest Vegetation" by the University of Florence for the Academic year 2015–2016. She is now contract Professor in Applied Botany at the University of Florence. Currently, she is also working in the management and protection of historic gardens using innovative approaches (remote sensing) to face stresses deriving from climate change. She was project manager of the LIFE MOTTLES project on ozone effect on forest ecosystems and responsible of the MOTTLES European network for the protection of forests from ozone. From 2014 to 2018, she was External Research Associate at the ForNaLab of Ghent University. She has completed numerous Postdoctoral Fellowships at various research institutes, e.g., for the "Monitoring the Ozone injuries on Forests" at the Institute of Sustainable Plant Protection, National Research Council of Italy (CNR) from 2016 to 2020, for the "Analyses of the Forest Biodiversity Related to Biotic and Abiotic Factors" from 2015 to 2016 and for the "Phytosanitary Monitoring of Sporadic Species under Biotic Threats" in 2013 at the Department of Agricultural and Environmental Production of Unifi. She is author of 29 articles published in SCI journals, 19 oral contributions, and 31 posters at national and international conferences.

Pierre Sicard PhD in Atmospheric Chemistry, is working on air pollution and climate change impacts on forests ecosystems to reduce the risk for plant ecosystems by using integrated assessment modeling, deposition model, epidemiological studies, and statistical and multivariate analysis. He is involved in numerous national and EU-funded projects as coordinator (e.g., FO3REST, AIRFRESH) or as Principal Investigator or on the steering committee (e.g., MOTTLES). He also has experience in assessment of air pollution impacts on human health (AirQ model) and developed the Aggregate Risk Index. He is very active in communication serving as Deputy Coordinator of the RG 8.04.00 "Air Pollution & Climate Change" under the International Union of Forest Research Organizations (IUFRO); involved as UNECE Expert Panel on Clean Air in Cities and active in the EU Clean Air Forum; member of the Editorial Board of journals (Environmental Research, Climate, Frontiers in Forests and Global Change); member of the scientific committee of meetings; and has published >60 papers and has a h-index of 26. He is also involved as Regional Expert Group on Climate in "Provence-Alpes-Côte d'Azur" region.

Editorial

SI: Air Pollution and Plant Ecosystems

Evgenios Agathokleous [1],*, Elisa Carrari [2] and Pierre Sicard [3]

[1] Institute of Ecology, School of Applied Meteorology,
    Nanjing University of Information Science and Technology (NUIST), Nanjing 210044, China
[2] National Research Council, Sesto Fiorentino, I-50019 Florence, Italy; elisa.carrari@ipsp.cnr.it
[3] ARGANS, Sophia Antipolis, 06410 Biot, France; pierre.sicard@acri-he.fr
* Correspondence: evgenios@nuist.edu.cn

Received: 30 July 2020; Accepted: 4 August 2020; Published: 9 August 2020

Abstract: Air pollution continues to be a serious issue for plant health and terrestrial ecosystems. In this issue of climate, some papers relevant to air pollution and its potential impacts on plant health and terrestrial ecosystems are collated. The papers provide some new insights and offer the opportunity to further advance the current understandings of air pollution and its linked impacts at different levels.

Keywords: air pollution; carbon dioxide; ethylenediurea; gross primary production; plant protection; tropospheric ozone; plant ecosystems

1. Introduction

Air pollution, and especially ground-level ozone (O_3) pollution, is a major issue for vegetation, challenging scientific and regulatory communities in a continuing effort to better understand air pollution and its impacts on vegetation [1–3]. Notable research progress has been observed over recent decades, highly advancing our understandings of air pollution spatiotemporal characteristics and trends [4–6] as well as air pollution effects on plants, from the molecular level to communities and ecosystems [1–3,7–9]. While air pollution spatiotemporal patterns and trends became clearer and air pollution impacts better understood, a vast array of these research programs suggests that there is still much to accomplish. Recognizing the need for more research in these topics, a Special Issue on "Air Pollution and Plant Ecosystems" is published in *Climate*. This Editorial presents the collective findings in the papers published in the *Climate* Special Issue "Air Pollution and Plant Ecosystems".

2. Special Issue Content

A total of 11 papers were submitted for potential publication within the Special Issue. Finally, six papers have been accepted for publication [10–15], translating to an acceptance rate of about 55%.

Fumagalli et al. [10] exposed grapevine (*Vitis vinifera*) to different O_3 levels over two growing seasons and revealed that high O_3 levels affected grapevine weight and yields. Their study suggests that wine quality can be affected by reduced polyphenols that can decrease the nutritional value of the agricultural product and induce a more aggressive taste to wine. This project provides evidence of potential O_3 impacts on the quality of grapes and wine, encouraging the implementation of further studies to examine the potential effects on animals consuming such products altered by O_3.

Tobita et al. [11] exposed *Fagus crenata* plants to ambient air, elevated CO_2 (550 µmol mol^{-1} CO_2), elevated O_3 (2 × ambient O_3), and elevated CO_2 combined with elevated O_3 during two growing seasons. They found that the total plant biomass and elongation of second-flush shoots were increased more by elevated CO_2 combined with elevated O_3, and less by elevated CO_2 alone. Both elevated O_3 and elevated CO_2, as single stresses, decreased biomass allocation to the roots. This research suggests that elevated concentrations of CO_2 mitigate the negative impacts of O_3 on net CO_2 assimilation.

Kitao et al. [12] analyzed the fate of absorbed light energy, including photosynthesis, photorespiration, and regulated and nonregulated nonphotochemical quenching, by using data from experiments studying the effects of nitrogen limitation and drought on Japanese white birch (*Betula platyphylla* var. *japonica*), as well as the effect of elevated O_3 on Japanese oak (*Quercus mongolica* var. *crispula*) and Konara oak (*Q. serrata*) under elevated CO_2 concentrations. The rate of regulated nonphotochemical quenching (J_{NPQ}) could compensate for decreases in the photosynthetic electron transport rate (J_{PSII}) under the different stresses. It was also found that even decreases in nonregulated nonphotochemical quenching (J_{NO}) occurred under limited nitrogen and elevated O_3, irrespective of CO_2 conditions. These may indicate a preconditioning adaptive response preparing plants to cope with predicted environmental challenges. The results of this study can be used as a platform upon which to base new studies directed at revealing whether elevated CO_2 may not affect the plant responses to environmental stresses in terms of susceptibility to photodamage occurring in different experimental systems.

Proietti et al. [13], considering the importance of soil water availability as a driver of vegetation productivity, analyzed the spatiotemporal variation of a proposed temperature vegetation wetness index as a proxy of soil moisture and evaluated its effect on gross primary production using 19 representative tree species in Europe over the time period 2000–2010. The Modified Temperature Vegetation Wetness Index (mTVWI) displayed minimum soil water availability in Southern Europe and maximum soil water availability in Northeastern Europe. Furthermore, gross primary productivity decreased from 20% to 80% by mTVWI, depending on the site, tree species, and meteorological conditions. This wetness index adds a new dimension in understanding the impacts of water deficit stress which often occurs in tandem with air pollution.

Pandey et al. [14] treated 11 Indian wheat (*Triticum aestivum*) cultivars grown in high ambient O_3 (twice the critical threshold for wheat yield) with the antiozonant chemical ethylenediurea (300 mg L^{-1}), and found a high variation in resource allocation strategies among cultivars. They found that plants treated with ethylenediurea (EDU) produced more grain yields and had a higher photosynthetic rate and stomatal conductance as well as lower lipid peroxidation. They also observed varied responses of superoxide dismutase activity, catalase activity, and oxidized and reduced glutathione content. Responses to EDU (or O_3 assuming the differences were due to ambient O_3) varied across cultivars and plant developmental stages and sites. Authors grouped cultivars into four groups according to their response strategies. This research provides useful information to better understand the determinants of tolerance/susceptibility of Indian wheat to ambient O_3.

El-Tahan [15] used data of the Total Ozone Column (TOC), yielded from the Atmospheric Infrared Sounder (AIRS) and the model Modern-Era Retrospective analysis for Research and Applications (MERRA). The long-term trend and the spatial distribution over Egypt are studied, and a comparison between both sources of TOC is made. According to the results, the spatial maps from AIRS could identify the location of both high and low concentrations of O_3. Conversely, spatial maps from MERRA-2 underestimated TOC and were not effective in capturing the variability identified by AIRS. The study concludes that the MERRA-2 dataset also underestimated the temporal TOC over Egypt compared to the AIRS dataset. Among others, this study indicates the need to construct TOC from numerical models, such as, for example, numerical weather research and forecasting models coupled with chemistry.

3. Conclusions

A total of six papers on a variety of topics related to air pollution and its impacts were published in this special issue, constituting an orchestrated collection for researchers, environmentalists, educators, and local or regional regulators interested in air pollution and its impacts on plant ecosystems. We wish you an enjoyable and informative reading.

Author Contributions: Conceptualization, writing—original draft preparation, writing—review and editing: E.A., E.C., and P.S. All authors have read and agreed to the published version of the manuscript.

Funding: This research received no external funding.

Acknowledgments: The Editors are grateful to all those who have submitted their works to this Special Issue. E.A. acknowledges multi-year financial support from The Startup Foundation for Introducing Talent of Nanjing University of Information Science & Technology (NUIST), Nanjing, China (No. 003080 to E.A.).

Conflicts of Interest: The authors declare no conflict of interest. The funders had no role in the design of the study; in the collection, analyses, or interpretation of data; in the writing of the manuscript; or in the decision to publish the results.

References

1. Sanz, J.; González-Fernández, I.; Elvira, S.; Muntifering, R.; Alonso, R.; Bermejo-Bermejo, V. Setting ozone critical levels for annual Mediterranean pasture species: Combined analysis of open-top chamber experiments. *Sci. Total Environ.* **2016**, *571*, 670–679. [CrossRef] [PubMed]

2. Harmens, H.; Mills, G.; Hayes, F.; Norris, D.A.; Sharps, K. Twenty eight years of ICP Vegetation: An overview of its activities. *Ann. Bot.* **2015**, *5*, 31–43.

3. Paoletti, E.; Feng, Z.; De Marco, A.; Hoshika, Y.; Harmens, H.; Agathokleous, E.; Domingos, M.; Mills, G.; Sicard, P.; Zhang, L.; et al. Challenges, gaps and opportunities in investigating the interactions of ozone pollution and plant ecosystems. *Sci. Total Environ.* **2020**, *709*, 136188. [CrossRef] [PubMed]

4. Schultz, M.G.; Schröder, S.; Lyapina, O.; Cooper, O.; Galbally, I.; Petropavlovskikh, I.; Von Schneidemesser, E.; Tanimoto, H.; Elshorbany, Y.; Naja, M.; et al. Tropospheric Ozone Assessment Report: Database and Metrics Data of Global Surface Ozone Observations. *Elem. Sci. Anth.* **2017**, *5*, 58. [CrossRef]

5. Mills, G.; Pleijel, H.; Malley, C.S.; Sinha, B.; Cooper, O.R.; Schultz, M.G.; Neufeld, H.S.; Simpson, D.; Sharps, K.; Feng, Z.; et al. Tropospheric ozone assessment report: Present-day tropospheric ozone distribution and trends relevant to vegetation. *Elementa* **2018**, *6*, 47. [CrossRef]

6. Chang, K.-L.; Petropavlovskikh, I.; Copper, O.R.; Schultz, M.G.; Wang, T. Regional trend analysis of surface ozone observations from monitoring networks in eastern North America, Europe and East Asia. *Elem. Sci. Anth.* **2017**, *5*, 50. [CrossRef]

7. Fuhrer, J.; Val Martin, M.; Mills, G.; Heald, C.L.; Harmens, H.; Hayes, F.; Sharps, K.; Bender, J.; Ashmore, M.R. Current and future ozone risks to global terrestrial biodiversity and ecosystem processes. *Ecol. Evol.* **2016**, *6*, 8785–8799. [CrossRef] [PubMed]

8. Ghosh, A.; Singh, A.A.; Agrawal, M.; Agrawal, S.B. *Ozone Toxicity and Remediation in Crop Plants*; Springer: Cham, Switzerland, 2018; pp. 129–169.

9. Izuta, T. *Air Pollution Impacts on Plants in East Asia*; Izuta, T., Ed.; Springer: Tokyo, Japan, 2017; ISBN 978-4-431-56436-2.

10. Fumagalli, I.; Cieslik, S.; De Marco, A.; Proietti, C.; Paoletti, E. Grapevine and Ozone: Uptake and Effects. *Climate* **2019**, *7*, 140. [CrossRef]

11. Tobita, H.; Komatsu, M.; Harayama, H.; Yazaki, K.; Kitaoka, S.; Kitao, M. Effects of Combined CO_2 and O_3 Exposures on Net CO_2 Assimilation and Biomass Allocation in Seedlings of the Late-Successional Fagus Crenata. *Climate* **2019**, *7*, 117. [CrossRef]

12. Kitao, M.; Tobita, H.; Kitaoka, S.; Harayama, H.; Yazaki, K.; Komatsu, M.; Agathokleous, E.; Koike, T. Light energy partitioning under various environmental stresses combined with elevated CO_2 in three deciduous broadleaf tree species in Japan. *Climate* **2019**, *7*, 79. [CrossRef]

13. Proietti, C.; Anav, A.; Vitale, M.; Fares, S.; Fornasier, M.F.; Screpanti, A.; Salvati, L.; Paoletti, E.; Sicard, P.; De Marco, A. A New Wetness Index to Evaluate the Soil Water Availability Influence on Gross Primary Production of European Forests. *Climate* **2019**, *7*, 42. [CrossRef]

14. Pandey, A.K.; Majumder, B.; Keski-Saari, S.; Kontunen-Soppela, S.; Pandey, V.; Oksanen, E.; Pandey, A.K.; Majumder, B.; Keski-Saari, S.; Kontunen-Soppela, S.; et al. High variation in resource allocation strategies among 11 Indian wheat (Triticum aestivum) cultivars growing in high ozone environment. *Climate* **2019**, *7*, 23. [CrossRef]

15. El-Tahan, M. Temporal and spatial ozone distribution over Egypt. *Climate* **2018**, *6*, 46. [CrossRef]

Article

Temporal and Spatial Ozone Distribution over Egypt

Muhammed El-Tahan

Aerospace Engineering Department, Cairo University, Cairo 12613, Egypt; muhammedsamireltahan@gmail.com

Received: 21 May 2018; Accepted: 27 May 2018; Published: 29 May 2018

Abstract: The long-term temporal trends and spatial distribution of Ozone (O_3) over Egypt is presented using monthly data from both the Atmospheric Infrared Sounder (AIRS) and the model Modern-Era Retrospective analysis for Research and Applications (MERRA) datasets. The twelve-year monthly record (2005–2016) of the Total Ozone Column (TOC) has a spatial resolution of $1 \times 1°$ from AIRS and $0.5 \times 0.625°$ from the MERRA-2 dataset. The average monthly, seasonal and interannual time series are analyzed for their temporal trends, while the seasonal average spatial distributions are compared. It was found that MERRA-2 underestimated AIRS measurements. Both AIRS and MERRA-2 have their minimum monthly averages of TOC in February 2013. The maximum monthly average TOC from AIRS is 321.48 DU in July 2012, while that from MERRA-2 is 303.48 in April 2011.

Keywords: AIRS; MERRA-2; ozone; trend; spatial and temporal O_3

1. Introduction

Ozone is a major greenhouse gas, thus, it plays an important role in both weather and climate, and its impact varies from global to regional scales [1,2]. While it represents only 0.0012% of the atmospheric composition [3], ozone acts as an absorber for the energetic particle from the solar ultraviolet radiation (UV), protecting the earth from harmful radiation [4,5], which has a harmful effect on human health particularly on the skin [6,7]. The observed increase in UV radiation at the earth's surface has been due to the decrease of amount of ozone at the stratospheric atmospheric layer [8–10], which is caused by photochemical losses related to anthropogenic reasons [11–14].

Therefore, the spatial and temporal variation of O_3 over global and regional domains has become an important research subject [14]. Global total column ozone concentration (which is referred to as ozone in the stratosphere) decreased a few percent between the 1970s and the start of this century [15]. In the stratosphere, ozone plays the role of a natural and beneficial screen in relation to the harmful effects of ultraviolet for the organic matter. In the troposphere, ozone is a secondary pollutant that is produced during the atmospheric photo-oxidation of volatile organic compounds under the presence of nitrogen oxides emitted, mainly, by anthropogenic activities, while surface ozone is considered to be the most damaging air pollutant in terms of adverse effects on human health, vegetation, crops and materials in Europe and may become worse in the future [16–20]. The total amount of ozone at any location on the globe is defined as the summation of all the ozone in the atmosphere directly above that location [21]. Evaluation of ozone profile and variability from different satellite data against in situ measurements (on board the NSF/NCAR Gulfstream-V aircraft during the Stratosphere-Troposphere Analyses of Regional Transport in 2008 (START08) experiment) from Atmospheric Infrared Sounder (AIRS), Infrared Atmospheric Sounding Interferometer (IASI), and the Ozone Monitoring Instrument (OMI) shows that the three satellite products have an acceptable capability to represent the variability of ozone in the upper troposphere and lower stratosphere. Statistical analyses revealed that the three satellite products captured 80% of the variability of ozone monitored in the aircraft data [22]. The comparison between the measured and reanalysis of Total Ozone Column (TOC) over Cairo city between 1979 and 2014 shows good agreement with (r = 0.91) the correlation coefficient [23]. The highest ozone measurements were recorded in the summer of 2007 over Cairo city [23]. The long-term variability of ozone from the main four ground urban observation stations in Cairo, Aswan, Matrouh and Hurgada shows that negative trend values in ozone are the dominant features during the period 1990–2014 at all stations [24]. Formation of Ozone over the greater Cairo was investigated based on two measurement campaigns. The first one was in 1990 and lasted for 3 weeks based on measurements from three different sites (Shoubra El-Kheima, Mokattam Hills and Helwan). The second one was in 1991 and lasted for 7 months from April to October based on one site at El-Kobba. It was found that ozone is produced over the industrial sites at north and center of Cairo and transported southward by the northerly winds [25]. Also, high average ozone levels were observed during the night in the spring and the summer [25]. The automatic station located 30 km south of Dekhla Oasis in Egypt in the Lybian desert (powered by a photovoltaic generator system to measure the vertical ozone flux) confirmed that high ozone fractions were recorded when northerly winds prevailed [26].

There is an interest in investigating the impact of air pollutants on agricultural crops, and this interest has focused on the long-term low-level effects of the main phytotoxic gases on crop production [27–32]. Periodic exposure to air pollutants may cause yield losses [27,33–39]. Apart from crop yield losses, changes in plant development and reduction in net growth can occur [40–42], as well as changes in crop quality [43–46]. Ozone is the main phytotoxic air pollutant in the Mediterranean area [47–54]. The high levels of O_3 over Egypt can cause significant decrease in the growth and the local varieties of crop plants [45]. On testing and selecting multiple sensitive and environmentally successful Egyptian bioindicator plants for ozone (O_3), four plant species (jute, clover, garden rocket and alfalfa) were found to be more sensitive to O_3 than the universally used O_3-bioindicator, tobacco Bel W3 [55].

The objective of the current work is to highlight the analysis of spatial-temporal of TOC over Egypt (23.7–36.2° N and 21.5–32.3° E), which could help understanding how Egypt contributes in this issue. In this study, TOC from the remote sensor AIRS and MERRA-2 datasets is used to generate the long-term trend and the spatial distribution over Egypt; a comparison between both sources of TOC is established. As far as the author is aware, research of this kind has not been conducted over Egypt before.

2. Methodology

In this study, TOC data from two different sources is introduced. Those sources are: AIRS and the model Modern-Era Retrospective analysis for Research and Applications version 2 (MERRA-2) dataset.

2.1. AIRS Sensor

AIRS was launched into orbit in 2002 aboard NASA's Aqua satellite; its primary goal is to support weather and climate research [56]. An innovative atmospheric sounding group of visible, infrared, and microwave sensors constitute the multispectral range of AIRS. The Level-3 data from AIRS (Daytime/Ascending) is used with a $1 \times 1°$ spatial resolution monthly gridded retrieval product (AIRX3STM) version (v006) [57].

AIRS has previously demonstrated its capacity to capture TOC. AIRS/Aqua level-3 daily gridded products (AIRX3STD) $1 \times 1°$ spatial resolution (version 5) data could successfully detect O_3 from large forest fires [58]. Benchmarking of AIRS (version 5)-retrieved ozone profiles with TOC has already been done by the World Ozone and Ultraviolet Radiation Data Center. The biases from the collected ozonesonde (O3SND) were less than 5% for both stratosphere and troposphere [22].

2.2. MERRA-2 Model Data

The MERRA-2 [59] dataset consists of worldwide meteorological variables hosted by NASA and generated by the Goddard Space Flight Center. This dataset is generated by the Goddard Earth Observing System Model, version 5 (GEOS-5). The spatial resolution is 0.625° in latitude and 0.5° in longitude (approx. 50 km). It replaces the original MERRA reanalysis [60]. Daily TOC and relative humidity (RH) from AIRS on NASA's Aqua satellite are used to identify the presence of SI over Rocky Mountain National Park in observational data, and to validate MERRA-2 reanalysis of TOC, since AIRS data are not assimilated by MERRA-2. AIRS is equipped to measure both meteorological variables and chemical profiles [61–63].

The MERRA-2 dataset is monthly averaged [64]. Since 2004, MERRA-2 has assimilated satellite retrievals of TOC from the Ozone Monitoring Instrument [33,65] and stratospheric O_3 profiles from the Microwave Limb Sounder [66–68]. TOC from the MERRA-2 dataset in the lower stratosphere has good representation and has proven the agreement with ozonesondes [69,70]. MERRA-2's total ozone agrees with Total Ozone Mapping Spectrometer (TOMS) data (1980–1993) very well, with less than 2% bias and less than 6% difference in standard deviation, which is close to the assumed observation error of 5% [70]. There is a good representation of the variability of stratospheric ozone in MERRA-2. The difference in standard deviations between the reanalysis data and the independent limb satellite data range from 11% for SAGE II in the lower stratosphere to less than 5% at 4.3 hPa [70].

3. Results

The temporal trends of TOC, which include monthly average, seasonal and interannual time series, are introduced in Section 3.1. The spatial distribution seasonal maps are presented in Section 3.2. Both time series and spatial maps of TOC are in Dobson units (DU).

3.1. Temporal Trend

Tropospheric ozone ground observations show that ozone has increased globally during the 20th century. Ozone records over Europe show that ozone has doubled between the 1950s and 2000 [71]. Daily total ozone observations from TOMS over Dundee city in Scotland in the period (1979–1992) show a significant negative trend [72]. Aircrafts measured significant upper tropospheric trends in one or more seasons above multiple locations the North Atlantic Ocean, the north-eastern USA, the Middle East, Europe, northern India, southern Japan and China. From 1990 to 2010, surface ozone trends have varied according to the region. Western Europe showed increasing ozone in the 1990s followed by a decrease since 2000 [71]. In eastern US, surface ozone has decreased strongly in the summer, remained unchanged in the spring, and it has increased in the winter; in other locations such as the in western US, ozone has increased the most in the spring. Surface ozone in East Asia is increasing [71]. In general, MERRA-2 underestimates TOC compared to AIRS over Egypt. Monthly average of TOC—in DU—over 12 years (2005–2016) is introduced in Figure 1 for both AIRS and MERRA-2. From AIRS, average TOC was 294.5 ± 16.5 DU. The max TOC was 321.14 DU on 1 July 2012 (286.2 DU from MERRA-2), while the minimum was 253.89 DU in February 2013. On the other hand, from MERRA-2, the average TOC was 277.8 ± 11.934 DU. The max TOC was 303.478 DU in April 2011 (317.01 DU from AIRS), while the minimum was 248.655 DU in February 2013. The basic statistics are summarized in Table 1.

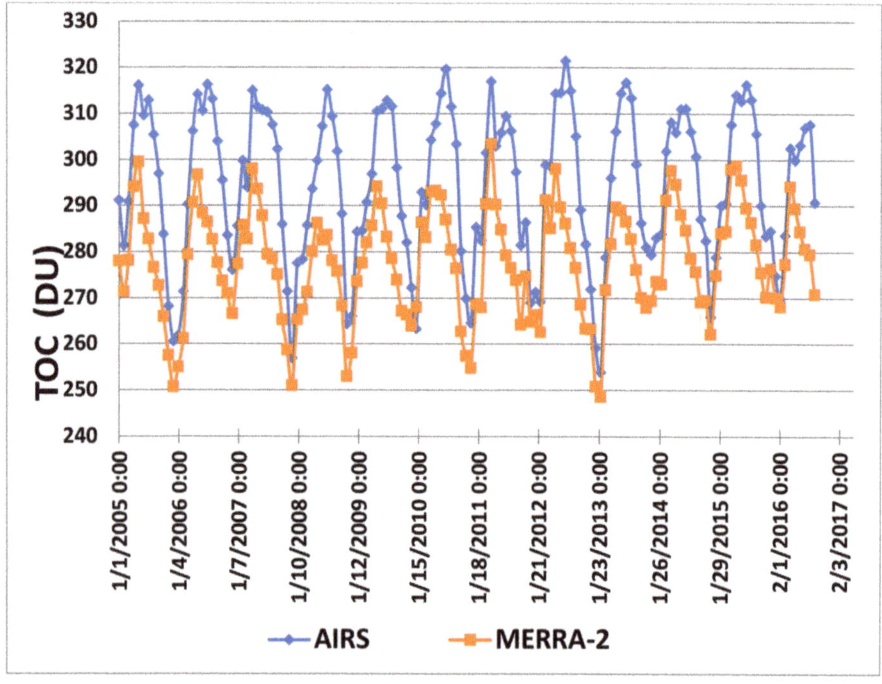

Figure 1. Temporal evolution of monthly average total ozone column over Egypt for the period (2005–2016) from AIRS and MERRA-2.

Table 1. Basic descriptive statistics for monthly average total ozone column over Egypt for the period (2005–2016) from AIRS and MERRA-2.

	AIRS	MERRA-2
Max	321.4 (July 2012)	303.4 (April 2011)
Min	253.8 (February 2013)	248.6 (February 2013)
Average	294.5	277.8
Variance	272.4	142.4
Std. Deviation	16.5	11.93

Figure 2 shows the probability density function (PDF) of TOC over Egypt of the 11-year dataset for both AIRS and MERRA-2, along with four tested distributions. Errors of fitting for the tested PDF using Kolmogorov–Smirnov goodness of fit are given in Table 2. The best fit between the given distribution and the hypothesized continuous distributions has the lowest error value. The best fit is ordered, and the order is shown in the parentheses. For AIRS, it is shown that that Weibull distribution has the best goodness of fit, with 0.08744 between the four tested distributions. Regarding the normal distribution case, the best goodness of fit, with 0.0473, was for the MERRA-2 data.

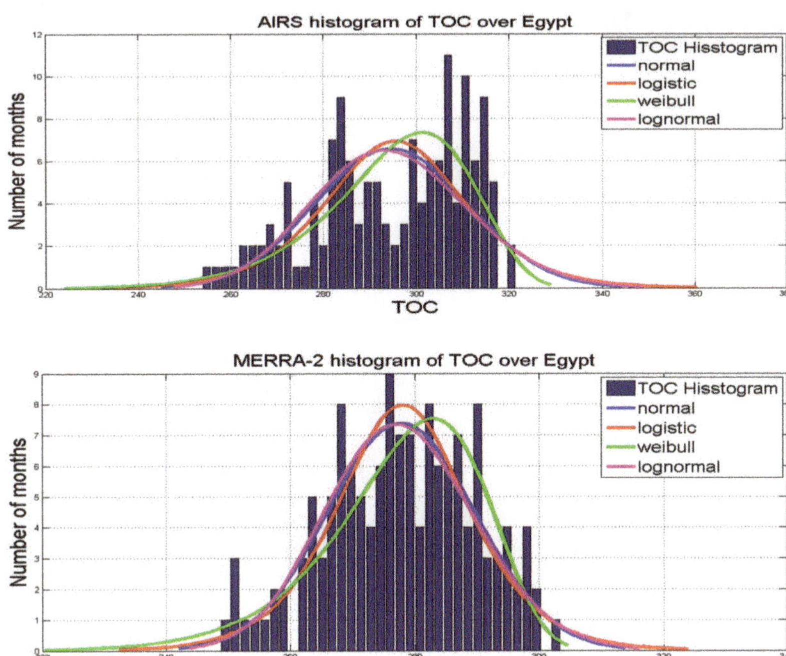

Figure 2. TOC histogram over Egypt for 12 years from AIRS (**upper**) and MERRA-2 (**lower**).

Table 2. Goodness of fit for both AIRS and MERRA-2 data over Egypt.

	Logistic	Lognormal	Normal	Weibull
AIRS	0.095 (2)	0.111 (4)	0.108 (3)	0.087 (1)
MERRA-2	0.048 (2)	0.053 (3)	0.047 (1)	0.055 (4)

Interannual monthly regional average time series (2006–2016) of TOC—in DU—over Egypt from both AIRS and MERRA-2 are shown in Figure 3. For AIRS, it is shown that July had the highest TOC values, while December had the lowest. It is noticeable from both AIRS and MERRA-2 that February had negative slope between 2010 and 2013, with an increase starting from 2014. February 2013 had the lowest TOC concentration in the eleven-year period of shown data with no clear reason behind this bottom point until now. On the other hand, for the MERRA-2 dataset, it can be seen that October 2011 had the highest TOC values, while February 2013 had the lowest. The basic statistics (maximum, minimum, average, variance and standard deviation) for interannual monthly average TOC from both AIRS and MERRA-2 are summarized in Tables 3 and 4. The corresponding year for max and min monthly average TOC is represented between the parentheses.

Table 3. Interannual monthly regional average statistics TOC for 12 years from AIRS. Numbers in parentheses represent the corresponding year.

	AIRS				
	Max.	**Min.**	**Average**	**Variance**	**Std. Deviation**
January	291.1 (2005)	259.1 (2013)	276.4	109.5	10.4
February	299.8 (2007)	253.8 (2013)	279.8	153.1	12.3
March	301.9 (2011)	278.8 (2013)	291.4	48.2	6.9
April	317.1 (2011)	293.7 (2008)	304.5	52.8	7.2
May	316.1 (2005)	299.8 (2008)	308.6	32.4	5.6
June	314.5 (2013)	303.3 (2016)	310.5	12.5	3.5
July	321.4 (2012)	307.1 (2016)	314.1	18.2	4.2
August	314.9 (2012)	305.4 (2005)	310.1	11.03	3.3
September	305.7 (2015)	290.8 (2016)	300.5	17.9	4.2
October	295.6 (2006)	280.2 (2010)	286.9	17.8	4.2
November	286.4 (2011)	264.4 (2008)	277.7	57.7	7.6
December	284.5 (2015)	256.9 (2007)	269.8	66.9	8.1

Table 4. Interannual monthly regional average statistics TOC for 12 years from MERRA-2. Numbers in parentheses represent the corresponding year.

	MERRA-2				
	Max.	**Min.**	**Average**	**Variance**	**Std. Deviation**
January	278.1 (2005)	250.8 (2013)	268.5	70.7	8.4
February	286.4 (2010)	248.6 (2013)	271.2	124.5	11.2
March	291.3 (2012)	271.3 (2013)	281.9	46.4	6.8
April	303.5 (2011)	280.1 (2008)	291.8	52.9	7.3
May	299.5 (2005)	286.2 (2008)	293.7	17.4	4.2
June	295.7 (2015)	282.6 (2008)	288.4	12.6	3.5
July	289.6 (2015)	279.3 (2011)	284.2	10.3	3.2
August	286.5 (2015)	276.6 (2011)	280.5	8.2	2.8
September	281.6 (2015)	270.9 (2016)	275.6	7.2	2.6
October	275.6 (2015)	262.7 (2010)	267.8	16.0	4.0
November	274.7 (2011)	253.5 (2016)	263.7	53.9	64.9
December	276.4 (2015)	250.7 (2005)	261.1	7.3	8.1

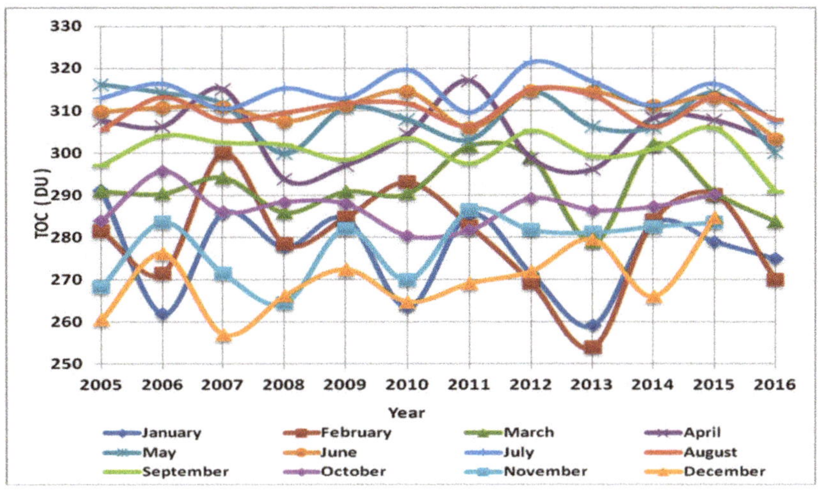

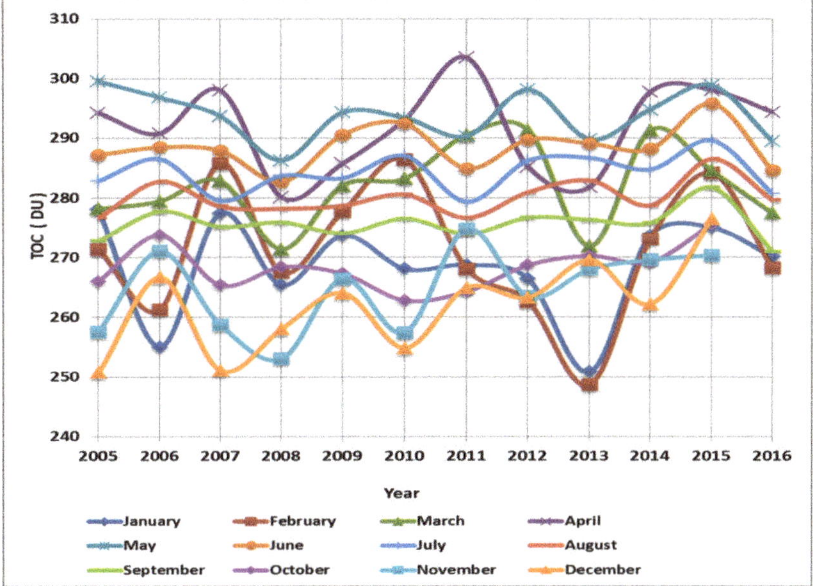

Figure 3. Interannual monthly average time series of TOC in Dobson Units (DU) over Egypt for the period (2005–2016) from AIRS (**upper**) and MERRA-2 (**lower**).

Seasonal average time series for TOC—in DU—is highlighted in Figure 4. It is shown that TOC in the summer (JJA) has the highest values, followed by the spring, then the fall season. The winter (DJF) has the lowest TOC. The highest TOC concentration in the summer, especially June, is comparable with previous studies conducted over Cairo by Y. Aboel Fetouh in 2013 [23]. The statistics of the seasonal average time series for TOC are introduced in Table 5. The number between the parentheses corresponds to each maximum or minimum seasonal TOC over Egypt.

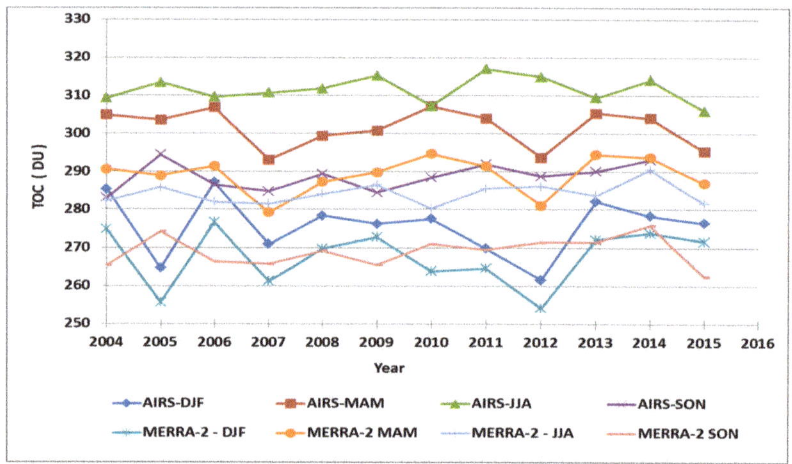

Figure 4. Seasonal monthly average time series of TOC in Dobson Unit (DU) over Egypt (2005–2016) from AIRS and MERRA-2.

As shown in Table 5, both AIRS and MERRA-2 sensors recorded maximum and minimum TOC measurements in the same years, 2006 and 2012, which occurred in the DJF season; however, this behavior is not repeated in any of the remaining seasons (MAM, JJA or SON), as shown by the table, i.e., the two sensors' maximum and minimum TOC measurements do not overlap in the same year again. AIRS provides the maximum TOC measurement for the DJF season with a value of 287.227 DU in year 2006, while the minimum was 261.67 DU in year 2012. On the other hand, MERRA-2 shows that the max. TOC in the DJF season was 276.644 DU and the minimum was 254.259 DU. Figure 5 shows the boxplots for the different seasons from both AIRS and MERRA-2. For AIRS, it is shown that the highest average season is JJA and the lowest is DJF, while for MERRA-2 the highest average season is MAM and the lowest is SON.

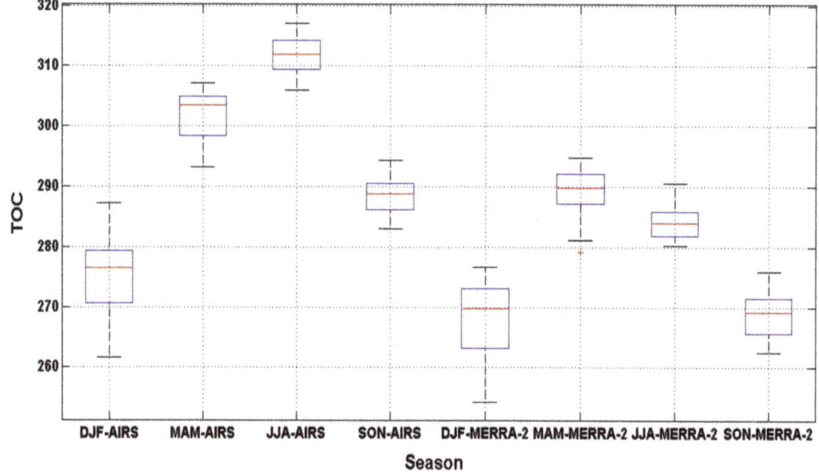

Figure 5. Boxplot seasonal monthly average of TOC (2005–2016) from AIRS and MERRA-2.

Table 5. Seasonal average time series statistics of TOC for 12 years from both AIRS and MERRA-2. The numbers in the parentheses represent the corresponding year.

	Max.	Min.	Average	Variance	Std. Deviation
AIRS					
DJF	287.2 (2006)	261.6 (2012)	275.7	60.3	7.7
MAM	307.2 (2011)	293.2 (2013)	302.1	23.64033	4.8
JJA	317.0 (2012)	306.1 (2016)	312.1	9.3971	3.1
SON	294.4 (2006)	283.1 (2005)	288.6	13.4	3.6
MERRA-2					
DJF	276.6 (2006)	254.2 (2012)	267.6	56.6	7.5
MAM	294.7 (2014)	279.2 (2008)	289.4	26.2	5.1
JJA	290.6 (2015)	280.3 (2011)	284.4	8.6	2.9
SON	275.8 (2015)	262.5 (2016)	269.6	12.7	3.5

The Mann-Kendall (MK) [73–75] test is a way to identify if a monotonic upward (increase) or downward (decrease) trend exists for a specific variable over time. This test is run on the seasonal monthly average TOC in Figure 4. The results of Mann-Kendall test for all seasonal monthly average TOCs for both AIRS and MIRRA-2 data found that there was no trend. Table 6 summarizes the analysis of the eight seasonal average time series from both AIRS and MERRA-2 under the Mann-Kendall test. There was no significant trend ($p > 0.05$) in the time series.

Table 6. Summary of the results of Mann-Kendall test for the eight seasonal monthly average TOCs over Egypt (2005–2016).

	p-Value	Significant Trend
DJF-AIRS	0.5	
MAM-AIRS	0.5	
JJA-AIRS	0.6	
SON-AIRS	0.19	No Trend
DJF-MERRA-2	0.27	
MAM-MERRA-2	0.32	
JJA-MERRA-2	0.38	
SON-MERRA-2	0.27	

3.2. Spatial Distribution

The spatial distribution of seasonal monthly averages of TOC over the 12-year study period for both AIRS and MERRA-2 is shown in Figure 6. The spatial maps show that MERRA-2 also underestimates TOC over Egypt. Once again, for AIRS, the winter map (DJF) reveals the lowest spatial distribution, while the summer map (JJA) reveals the highest spatial distribution, confirming the time series in Figure 3.

The right panel from Figure 6 shows also that the north coast of Egypt has the highest TOC in the spring (higher than the summer), which is associated with the main and secondary Mediterranean cyclones [24], while the higher concentration in summer is due to trans-boundary transportation of ozone from Europe [23,76]. Those regions include the Nile Delta (D1) and the Qattara Depression and area (D2). Anthropogenic activities and emissions in urban and industrial areas have had an impact on ozone production and concentration [77]. The Nile Delta (D1) region has high levels of anthropogenic activities. Meanwhile, Qattara Depression is considered an active dust source and contains both clay and sand. In addition, several studies have suggested a link between ozone and fine particulate matter (PM) concentrations [76–79]. This may be the reason for high ozone in the D2 domain. On the other hand, the left panel from Figure 6 shows that MERRA-2 does not have the capability to capture the same variability of TOC over Egypt as AIRS does, although it has higher resolution.

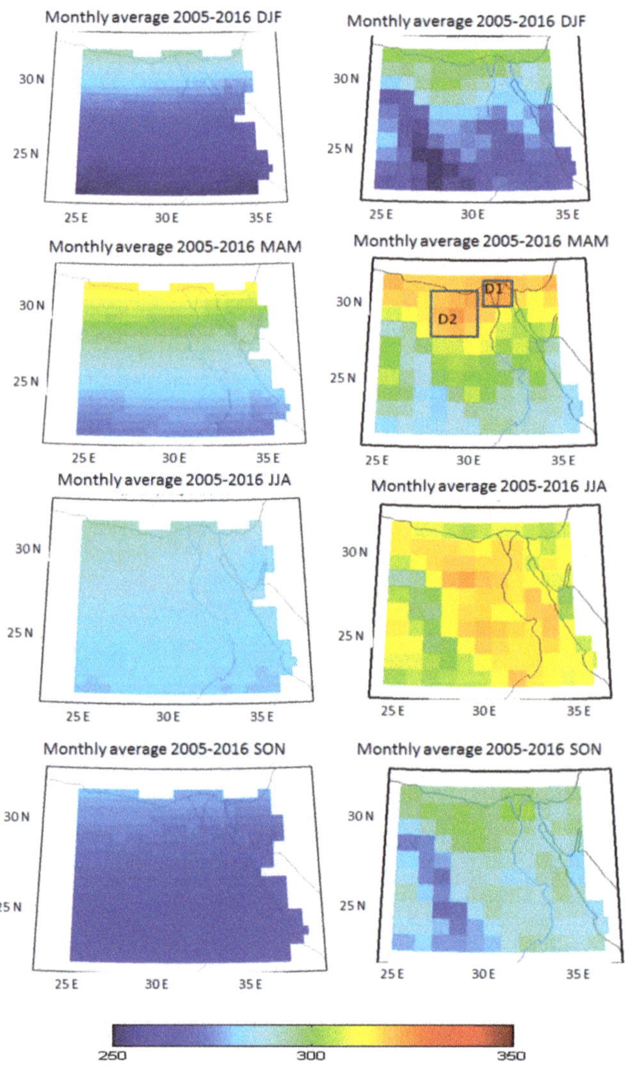

Figure 6. Seasonal average of TOC in (DU) over Egypt for winter (DJF), spring (MAM), summer (JJA) and fall (SON), for the period (2005–2016). Two domains, D1 and D2, are shown for further analysis on MAM map for MERRA-2 (**left panel**) and AIRS (**right panel**).

4. Conclusions and Future Works

The AIRS and MERRA-2 datasets provide observations for the TOC that allow the tracking and monitoring of its concentration and its temporal and spatial variability over Egypt. Monthly and seasonal average time series show that summer months feature the highest TOC. Comparing the two sensors with regard to TOC measurement, it was found that the spatial maps from AIRS over Egypt can identify the location of high and low concentrations of ozone, while spatial maps from MERRA-2 underestimate TOC and do not have the capability to capture the same variability shown by AIRS. It was concluded that the MERRA-2 dataset also underestimates the temporal TOC over Egypt in comparison to the AIRS data set. The statistical quantities for the 12-year temporal trend over Egypt are summarized.

Future work shall include investigation of the construction of TOC from numerical models, such as numerical weather research and forecasting models coupled with chemistry (WRF-chem). The sensitivity of TOC to different chemical mechanisms should be investigated. In addition to a study on the correlation between TOC from AIRS and aerosol optical depth from (MODIS and numerical weather research and forecasting coupled with chemistry (WRF-chem) model) shall be established.

Conflicts of Interest: The author declares no conflict of interest.

References

1. Kiehl, J.T.; Schneider, T.L.; Portmann, R.W.; Solomon, S. Climate forcing due to tropospheric and stratospheric ozone. *J. Geophys. Res.* **1999**, *104*, 31239–31254. [CrossRef]
2. Rex, M.; von der Salawitch, R.J.P.; Gathen, N.R.; Harris, P.; Chipperfield, M.; Naujokat, B. Arctic ozone loss and climate change. *Geophys. Res. Lett.* **2004**, *31*, L04116. [CrossRef]
3. Kerr, J.B.; McElroy, C.T. Total ozone measurements made with the Brewer ozone spectrophotometer during STOIC. *J. Geophys. Res.* **1989**, *100*, 9225–9230. [CrossRef]
4. Varotsos, C.A.; Chronpoulos, G.J.; Katsiki, S.; Sakellariou, N.K. Further evidence of the role of air-pollution on solar ultraviolet-radiation reaching the ground. *Int. J. Remote Sens.* **1995**, *16*, 1883–1886. [CrossRef]
5. Alexandris, D.; Varotsos, C.; Kondratyev, K.Y.; Chronopoulos, G. On the altitude dependence of solar effective UV. *Phys. Chem. Earth Part C Sol. Terr. Planet. Sci.* **1999**, *24*, 515–517. [CrossRef]
6. World Health Organization (WHO). *Protection Against Exposure to Ultraviolet Radiation*; Technical Report WHO/EHG #17; WHO: Geneva, Switzerland, 1995.
7. Kondratyev, K.Y.; Varotsos, C.A. Global total ozone dynamics—Impact on surface solar ultraviolet radiation variability and ecosystems. *Environ. Sci. Pollut. Res.* **1996**, *3*, 205. [CrossRef] [PubMed]
8. Fioletov, V.E.; McArthur, L.; Kerr, J.E.; Wardle, D.I. Long-term variations of UV-Birradiance over Canada estimated from Brewer observations and derived from ozone and pyranometer measurements. *J. Geophys. Res.* **2001**, *106*, 2307–2309. [CrossRef]
9. Frederick, J.E.; Manner, V.W.; Booth, C.R. Inter-annual variability in solar ultraviolet irradiance over decadal time scales at latitude 55 south. *Photochem. Photobiol.* **2001**, *74*, 771–779. [CrossRef]
10. Molina, M.J.; Rowland, F.S. Stratospheric sink for chlorofluoromethanes: Chlorine atom-catalysed destruction of ozone. *Nature* **1974**, *249*, 810–812. [CrossRef]
11. Farman, J.C.; Gardiner, B.G.; Shanklin, J.D. Large losses of total ozone in Antarctica reveal seasonal ClOx/NOx interaction. *Nature* **1985**, *315*, 207–210. [CrossRef]
12. Stolarski, R.S.; Krueger, A.J.; Schoeberl, M.R.; McPeters, R.D.; Newman, P.A.; Alpert, J.C. Nimbus 7 satellite measurements of the springtime Antarctic ozone decrease. *Nature* **1986**, *322*, 808–811. [CrossRef]
13. Harris, N.R.P.; Ancellet, J.; Bishop, L.; Hofmann, D.J.; Kerr, J.B.; McPeters, R.D.; Prendez, W.J.; Randel, J.; Staehelin, B.H.; Subbaraya, A. Trends in stratospheric and free tropospheric ozone. *J. Geophys. Res.* **1997**, *102*, 1571–1590. [CrossRef]
14. World Meteorological Organization (WMO). *Scientific Assessment of Ozone Depletion: Global Ozone Research and Monitoring Project*; Technical Report 50; WMO: Geneva, Switzerland, 2006.
15. World Meteorological Organization (WMO). *Scientific Assessment of Ozone Depletion: Global Ozone Research and Monitoring Project*; Report No. 52; WMO: Geneva, Switzerland, 2011.

16. Ochoa-Hueso, R.; Munzi, S.; Alonso, R.; Arróniz-Crespo, M.; Avila, A.; Bermejo, V.; Bobbink, R.; Branquinho, C.; Concostrina-Zubiri, L.; Cruz, C. Ecological Impacts of Atmospheric Pollution and Interactions with Climate Change in Terrestrial Ecosystems of the Mediterranean Basin: Current Research and Future Directions. *Environ. Pollut.* **2017**, *227*, 194–206. [CrossRef] [PubMed]

17. Paoletti, E.; de Marco, A.; Beddows, D.C.S.; Harrison, R.M.; Manning, W.J. Ozone levels in European and USA cities are increasing more than at rural sites, while peak values are decreasing. *Environ. Pollut.* **2014**, *192*, 295–299. [CrossRef] [PubMed]

18. Proietti, C.; Anav, A.; de Marco, A.; Sicard, P.; Vitalea, M. A multi-sites analysis on the ozone effects on Gross Primary Production of European forests. *Sci. Total Environ.* **2016**, *556*, 1–11. [CrossRef] [PubMed]

19. Sicard, P.; Alessandro, A.; Alessandra, D.; Elena, P. Projected global tropospheric ozone impacts on vegetation under different emission and climate scenarios. *Atmos. Chem. Phys. Discuss.* **2017**, 1–34. [CrossRef]

20. Lefohn, A.S.; Malley, C.S.; Smith, L.; Wells, B.; Hazucha, M.; Simon, H.; Naik, V.; Mills, G.; Schultz, M.G.; Paoletti, E.; et al. Tropospheric Ozone Assessment Report: Global ozone metrics for climate change, human health, and crop/ecosystem research. *Elem. Sci. Anth.* **2018**, *6*, 28. [CrossRef]

21. David, W.F.; Michaela, I. Twenty Questions and Answers about the Ozone Layer. 2014. Available online: http://www.atmos.umd.edu/~rjs/class/spr2017/readings/WMO_Ozone_2010_QAs.pdf (accessed on 29 May 2018).

22. Pittman, J.V.; Pan, L.L.; Wei, J.C.; Irion, F.W.; Liu, X.; Maddy, E.S.; Barnet, C.D.; Chance, K.; Gao, R.-S. Evaluation of AIRS, IASI, and OMI ozone profile retrievals in the extratropical tropopause region using in situ aircraft measurements. *J. Geophys. Res.* **2009**, *114*, D24109. [CrossRef]

23. Fetouh, Y.A.; El Askary, H.; El Raey, M.; Allali, M.; Sprigg, W.A.; Kafatos, M. Annual Patterns of Atmospheric Pollutions and Episodes over Cairo Egypt. *Adv. Meteorol.* **2013**, *2013*, 984853. [CrossRef]

24. Badawy, A.; Basset, H.A.; Eid, M. Spatial and Temporal Variations of Total Column Ozone over Egypt. *J. Earth Atmos. Sci.* **2017**, *2*, 1–16.

25. Güsten, H.; Heinrich, G.; Weppner, J.; Abdel-Aal, M.M.; Abdel-Hay, F.A.; Ramadan, A.B.; Tawfik, F.S.; Ahmed, D.M.; Hassan, G.K.Y.; Cvitašd, T.; et al. Ozone formation in the greater Cairo area. *Sci. Total Environ.* **1994**, *155*, 285–295. [CrossRef]

26. Güsten, H.; Heinrich, G.; Monnich, D.; Sprung, D.; Weppner, J.; BakrRamadan, A.; El-Din, M.R.M.E.; Ahmed, D.M.; Hassan, G.K.Y. On-line measurements of ozone surface fluxes: Part II; surface-level ozone fluxes onto the Sahara desert. *Atmos. Environ.* **1996**, *30*, 911–918. [CrossRef]

27. Weigel, H.J.; Adaros, G.; Jäger, H.J. An open top chamber study with filtered and non-filtered air to evaluate effects of air pollutants on crops. *Environ. Pollut.* **1987**, *47*, 231–244. [CrossRef]

28. De Temmerman, L.; Vandermeiren, K.; Guns, M. Effects of air filtration on spring wheat grown in open-top chambers at a rural site. I. Effect on growth, yield and dry matter portioning. *Environ. Pollut.* **1992**, *77*, 1–5. [CrossRef]

29. Schenone, G.; Botteschi, G.; Fumagali, I.; Montinaro, F. Effects of ambient air pollution in opentop chambers on bean (*Phaseolus vulgaris* L.) I. Effects on growth and yield. *New Phytol.* **1992**, *122*, 689–697. [CrossRef]

30. Schenone, G.; Fumagali, I.; Mignanego, L.; Montinaro, F.; Soldatini, G.F. Effects of ambient air pollution in open-top chambers on bean (*Phaseolus vulgaris* L.). II. Effects on photosynthesis and stomatal conductance. *New Phytol.* **1994**, *126*, 309–331. [CrossRef]

31. Hassan, I.A. Physiological and biochemical response of potato (*Solanum tuberosum* L. Cv. Kara) to O_3 and antioxidant chemicals: Possible roles of antioxidant enzymes. *Ann. Appl. Biol.* **2006**, *146*, 134–142. [CrossRef]

32. Hassan, I.A. Interactive effects of O_3 and CO_2 on growth, physiology of potato (*Solanum tuberosum* L.). *World J. Environ. Sustain. Dev.* **2010**, *7*, 1–12.

33. Heagle, A.S.; Philbeck, R.B.; Rogers, H.H.; Letchworth, M.B. Dispersing and monitoring O_3 in open-top field chambers for plant-effects studies. *Phytopathology* **1979**, *69*, 15–20. [CrossRef]

34. Pande, P.C.; Mansfield, T.A. Responses of spring barely to SO_2 and NO_2 pollution. *Environ. Pollut.* **1985**, *38*, 87–97. [CrossRef]

35. Schenone, G.; Lorenzini, G. Effects of regional air pollution on crops in Italy. *Agric. Ecosyst. Environ.* **1992**, *38*, 55–66. [CrossRef]

36. Ali, E.A. Damage to plants due to industrial pollution and their use as bioindicators in Egypt. *Environ. Pollut.* **1993**, *81*, 251–255. [CrossRef]

37. Dizengremel, P.; Le Thiec, D.; Bagard, M.; Jolivet, Y. Ozone risk assessment for plants: Central role of metabolism-dependent changes in reducing power. *Environ. Pollut.* **2008**, *156*, 11–15. [CrossRef] [PubMed]

38. Dizengremel, P.; Le Thiec, D.; Hasenfratz-Sauderm, M.P.; Vaultier, M.N.; Bagard, M.; Jolivet, Y. Metabolic-dependent changes in plant cell redox power after ozone exposure. *Plant Biol.* **2009**, *11*, 35–42. [CrossRef] [PubMed]

39. Dizengremel, P.; Vaultier, M.N.; Le Thiec, D.; Cabane, M.; Bagard, M.; Gerant, D.; Joëlle, G.; Allah Dghim, A.; Richet, N.; Afif, D.; et al. Phosphoenolpyruvate is at the crossroads of leaf metabolic responses to ozone stress. *New Phytol.* **2012**, *195*, 512–517. [CrossRef] [PubMed]

40. Gould, R.P.; Mansfield, T.A. Effects of SO_2 and NO_2 on growth and translocation in winter wheat. *J. Exp. Bot.* **1988**, *39*, 389–399. [CrossRef]

41. Hassan, I.A. Interactive effects of salinity and ozone pollution on photosynthesis, stomatal conductance, growth, and assimilate partitioning of wheat (*Triticum aestivum* L.). *Photosynthetica* **2004**, *42*, 111–118. [CrossRef]

42. Hatata, M.; Badr, R.; Ibrahim, M.; Hassan, I.A. Effects of O_3 and CO_2 on growth, yield and physiology of wheat (Triticum aestivum L.). *Curr. World Environ.* **2013**, *8*, 421–429.

43. Peleijel, H.; Skärby, L.; Wallin, G.; Sellden, G. Effects of grain quality of spring wheat exposed to O_3 in OTCs. In *The European Communities on Open-Top Chambers, Results on Agricultural Crops 1987–1988*; Air Pollution Rept.; Bonte, J., Mathy, P., Eds.; CEC: Brussels, Belgium, 1989; pp. 73–89.

44. Fuhrer, J.; Lehnherr, B.; Tschannen, W.; Moeri, P.B.; Shariat-Madari, H. Effects of O_3 on the grain composition of spring wheat grown in open-top field chambers. *Environ. Pollut.* **1990**, *65*, 181–192. [CrossRef]

45. Vandermeiren, K.; De Temmerman, L.; Staquet, A.; Baeten, H. Effects of air filtration on spring wheat grown in open-top field chambers at a rural site. II. Effects on mineral portioning, sulphur and nitrogen metabolism and on grain quality. *Environ. Pollut.* **1992**, *77*, 7–14. [CrossRef]

46. Hassan, I.A. Air pollution in Alexandria region, Egypt. II: Effect of regional air pollution on growth and yield of bean (Phaseolus vulgaris L. cv. Giza 6). In Proceedings of the 6th Egyptian Botanical Conference, Cairo, Egypt, 24–26 November 1998; Volume 3, pp. 469–479.

47. Lorenzini, G.; Nali, C.; Panicucci, A. Surface ozone in Pisa (Italy): A six-year study. *Atmos. Environ.* **1994**, *28*, 3155–3164. [CrossRef]

48. Hassan, I.A. Air pollution in Alexandria region, Egypt. I. An investigation of air quality. *Int. J. Environ. Educ. Inf.* **1999**, *18*, 67–78.

49. Hassan, I.A.; Ashmore, M.R.; Bell, J.N.B. Effects of O_3 on the stomatal behaviour of Egyptian verities of radish (*Raphanus sativus* L. cv. Baladey) and turnip (*Brassica rapa* L. cv. Sultani). *New Phytol.* **1994**, *128*, 243–249. [CrossRef]

50. Hassan, I.A.; Ashmore, M.R.; Bell, J.N.B. Effect of ozone on radish and turnip under Egyptian field conditions. *Environ. Pollut.* **1995**, *89*, 107–114. [CrossRef]

51. Hassan, I.A.; Basahi, J.M.; Ismail, I.; Habbebullah, T. Spatial distribution and temporal variation in ambient ozone and its associated NOx in the atmosphere of Jeddah City. *Saudi Arab. Aerosol. Air Qual.* **2013**, *13*, 1712–1722. [CrossRef]

52. Velissariou, D.; Gimeno, B.S.; Badiani, M.; Fumagalli, I.; Davison, A.W. Records of O_3 visible injury in the ECE Mediterranean region. In *Critical Levels for Ozone in Europe: Testing and Finalizing the Concept*; UN-ECE Workshop Report; Kärelampi, L., Skärby, L., Eds.; University of Kuopio: Kuopio, Finland, 1996; pp. 343–350.

53. Gimeno, B.S.; Bermejo, V.; Reinert, R.A.; Zheng, Y.; Barnes, J.D. Adverse effects of ambient ozone on watermelon yield and physiology at a rural site in Eastern Spain. *New Phytol.* **1999**, *144*, 245–260. [CrossRef]

54. Pellegrini, E.; Carucci, M.G.; Campanella, A.; Lorenzini, G.; Nali, C. Ozone stress in Melissa officinalis plants assessed by photosynthetic function. *Environ. Exp. Bot.* **2011**, *73*, 94–101. [CrossRef]

55. Madkour, S.A.; Laurence, J.A. Egyptian plant species as new ozone indicators. *Environ. Pollut.* **2002**, *120*, 339–353. [CrossRef]

56. *AIRS/AMSU/HSB Version 6 Data Release User Guide, Jet Propulsion Laboratory*; Version 1.2.1.; California Institute of Technology: Pasadena, CA, USA, 2005.

57. *AIRS Science Team/Joao TexeiraAIRS/Aqua L3 Monthly Standard Physical Retrieval (AIRS+AMSU) 1 Degree × 1 Degree V006*; Goddard Earth Sciences Data and Information Services Center (GES DISC): Greenbelt, MD, USA, 2013. [CrossRef]

58. Rajab, J.M.; MatJafri, M.Z.; Lim, H.S.; Abdullah, K. Daily distribution Map of Ozone (O_3) from AIRS over Southeast Asia. *Energy Res. J.* **2010**, *2*, 158–164. [CrossRef]

59. Bosilovich, M.; Akella, S.; Coy, L.; Cullather, R.; Draper, C.; Gelaro, R.; Kovach, R.; Liu, Q.; Molod, A.; Norris, P.; et al. *MERRA-2: Initial Evaluation of the Climate*; NASA/TM–2015–104606; National Aeronautics and Space Administration (NASA): Greenbelt, MD, USA, 2015; Volume 43, p. 139.

60. Rienecker, M.M.; Suarez, M.J.; Gelaro, R.; Todling, R.; Bacmeister, J.; Liu, E.; Bosilovich, M.G.; Schubert, S.D.; Takacs, L.; Kim, G. MERRA—NASA's Modern-Era Retrospective Analysis for Research and Applications. *J. Clim.* **2011**, *24*, 3624–3648. [CrossRef]

61. Susskind, J.; Barnet, C.; Blaisdell, J.; Iredell, L.; Keita, F.; Kouvaris, L.; Molnar, G.; Chahine, M. Accuracy of geophysical parameters derived from Atmospheric Infrared Sounder/Advanced Microwave Sounding Unit as a function of fractional cloud cover. *J. Geophys. Res.* **2006**, *111*, D09S17. [CrossRef]

62. Chahine, M.T.; Pagano, T.S.; Aumann, H.H.; Atlas, R.; Barnet, C.; Blaisdell, J.; Chen, L.; Divakarla, M.; Fetzer, E.J.; Zhou, L.; et al. AIRS: Improving weather forecasting and providing new data on greenhouse gases. *Bull. Am. Meteorol. Soc.* **2006**, *87*, 911–926. [CrossRef]

63. Aumann, H.H.; Chahine, M.T.; Gautier, C.; Goldberg, M.D.; Kalnay, E.; McMillin, L.M.; Revercomb, H.; Rosenkranz, P.W.; Smith, W.L.; Susskind, J.; et al. AIRS/AMSU/HSB on the aqua mission: Design, science objectives, data products, and processing systems. *IEEE Trans. Geosci. Remote Sens.* **2003**, *41*, 253–264. [CrossRef]

64. Bosilovich, M.G.; Lucchesi, R.; Suarez, M. MERRA-2: File Specification GMAO Office Note No. 9 (Version 1.1). 2016. Available online: http://gmao.gsfc.nasa.gov/pubs/docs/Bosilovich785.pdf (accessed on 20 June 2016).

65. Levelt, P.F.; van den Oord, G.H.J.; Dobber, M.R.; Malkki, A.; Visser, H.; de Vries, J.; Stammes, P.; Lundell, J.O.V.; Saari, H. The Ozone Monitoring Instrument. *IEEE Trans. Geosci. Remote Sens.* **2006**, *44*, 1093–1101. [CrossRef]

66. Waters, J.W.; Froidevaux, L.; Harwood, R.S.; Jarnot, R.F.; Pickett, H.M.; Read, W.G.; Siegel, P.H.; Cofield, R.E.; Filipiak, M.J.; Walch, M.J.; et al. The Earth Observing System Microwave Limb Sounder (EOS MLS) on the Aura satellite. *IEEE Trans. Geosci. Remote Sens.* **2006**, *44*, 1075–1092. [CrossRef]

67. Gelaro, R.; McCarty, W.; Suarez, M.J.; Todling, R.; Molod, A.; Takacs, L.; Bosilovich, M.G.; Schubert, S.D.; Takacs, L.; Zhao, B.; et al. The Modern-Era Retrospective Analysis for Research and Applications, Version 2 (MERRA-2). *J. Clim.* **2017**, *30*, 5419–5454. [CrossRef]

68. McCarty, W.; Coy, L.; Gelaro, R.; Huang, A.; Merkova, D.; Smith, E.; Meta, S.; Wargan, K. *MERRA-2 Input Observations: Summary and Assessment*; NASA/TM–2016–104606; NASA: Greenbelt, MD, USA, 2016; Volume 46, p. 64.

69. Wargan, K.; Labow, G.; Frith, S.; Pawson, S.; Livesey, N.; Partyka, G. Evaluation of the ozone fields in NASA's MERRA-2 reanalysis. *J. Clim.* **2017**, *30*, 2961–2988. [CrossRef] [PubMed]

70. Wargan, K.; Pawson, S.; Olsen, M.A.; Witte, J.C.; Douglass, A.R.; Ziemke, J.R.; Strahan, S.E.; Nielsen, J.E. The global structure of uppertroposphere-lower stratosphere ozone in GEOS-5: A multiyear assimilation of EOS Aura data. *J. Geophys. Res. Atmos.* **2015**, *120*, 2013–2036. [CrossRef]

71. Cooper, O.R.; Parrish, D.D.; Ziemke, J.; Balashov, N.V.; Cupeiro, M.; Galbally, I.E.; Gilge, S.; Horowitz, L.; Jensen, N.R.; Lamarque, J.-F.; et al. Global distribution and trends of tropospheric ozone: An observation-based review. *Elem. Sci. Anth.* **2014**, *2*, 29. [CrossRef]

72. Cracknell, A.P.; Varotsos, C.A. Ozone depletion over Scotland as derived from Nimbus-7 TOMS measurements. *Int. J. Remote Sens.* **1994**, *15*, 2659–2668. [CrossRef]

73. Gilbert, R.O. *Statistical Methods for Environmental Pollution Monitoring*; Wiley: Hoboken, NJ, USA, 1987.

74. Mann, H.B. Non-parametric tests against trend. *Econometrica* **1945**, *13*, 163–171. [CrossRef]

75. Kendall, M.G. *Rank Correlation Methods*, 4th ed.; Charles Griffin: London, UK, 1975.

76. Duncan, B.N.; West, J.J.; Yoshida, Y.; Fiore, A.M.; Ziemke, J.R. The influence of European pollution on ozone in the Near East and northern Africa. *Atmos. Chem. Phys.* **2008**, *8*, 2267–2283. [CrossRef]

77. Berntsen, T.K.; Isaksen, I.S.A.; Myhre, G.; Fuglestvedt, J.S.; Stordal, F.; Larsen, T.A.; Freckleton, R.S. Shine Effects of anthropogenic emissions on tropospheric ozone and its radiative forcing. *J. Geophys. Res.* **1997**, *102*, 28101–28126. [CrossRef]

78. Schwartz, J. Air pollution and hospital admissions for respiratory disease. *Epidemiology* **1996**, *7*, 20–28. [CrossRef] [PubMed]
79. Dockery, D.W.; Pope, C.A.; Xu, X.P.; Spengler, J.D.; Ware, J.H.; Fay, M.E.; Ferris, B.G.; Speizer, F. An association between air-pollution and mortality in 6 United States cities. *N. Engl. J. Med.* **1993**, *329*, 1753–1759. [CrossRef] [PubMed]

Article

High Variation in Resource Allocation Strategies among 11 Indian Wheat (*Triticum aestivum*) Cultivars Growing in High Ozone Environment

Ashutosh K. Pandey [1,2,†], Baisakhi Majumder [2], Sarita Keski-Saari [1], Sari Kontunen-Soppela [1], Vivek Pandey [2,*] and Elina Oksanen [1]

[1] Department of Environmental and Biological Sciences, University of Eastern Finland, POB 111, 80101 Joensuu, Finland; aashu.p20@gmail.com (A.K.P.); sarita.keski-saari@uef.fi (S.K.-S.); sari.kontunen-soppela@uef.fi (S.K.-S.); elina.oksanen@uef.fi (E.O.)
[2] Plant Ecology and Environmental Science, National Botanical Research Institute (CSIR-NBRI), Lucknow 226001, India; baisakhi.nbri12@gmail.com
* Correspondence: v.pandey@nbri.res.in; Tel.: +91-94506-57233
† Institute of Technology, University of Tartu, 50411 Tartu, Estonia (Current address).

Received: 26 December 2018; Accepted: 23 January 2019; Published: 28 January 2019

Abstract: Eleven local cultivars of wheat (*Triticum aestivum*) were chosen to study the effect of ambient ozone (O_3) concentration in the Indo-Gangetic Plains (IGP) of India at two high-ozone experimental sites by using 300 ppm of Ethylenediurea (EDU) as a chemical protectant against O_3. The O_3 level was more than double the critical threshold reported for wheat grain production (AOT40 8.66 ppm h). EDU-grown plants had higher grain yield, biomass, stomatal conductance and photosynthesis, less lipid peroxidation, changes in superoxide dismutase and catalase activities, changes in content of oxidized and reduced glutathione compared to non-EDU plants, thus indicating the severity of O_3 induced productivity loss. Based on the yield at two different growing sites, the cultivars could be addressed in four response groups: (a) generally well-adapted cultivars (above-average yield); (b) poorly-adapted (below-average yield); (c) adapted to low-yield environment (below-average yield); and (d) sensitive cultivars (adapted to high-yield environment). EDU responses were dependent on the cultivar, the developmental phase (vegetative, flowering and harvest) and the experimental site.

Keywords: cultivars; EDU (ethylenediurea); grain yield; India; ozone; wheat

---

1. Introduction

Tropospheric ozone (O_3) is a phytotoxic pollutant causing substantial damage to agricultural production and food security [1–7]. O_3-induced loss in plant productivity has been estimated to range between 14 and 26 billion US\$ on the global scale [8]. Modelling studies suggest a further increase in O_3 concentrations, especially in the East and South Asian regions due to the increased O_3 precursor emissions that result from high population density, rapid industrial growth and favorable climatic conditions [9–11]. Tropospheric O_3 concentration has increased in India and China by 20% and 13%, respectively, from the year 1999 to 2013 [12]. Recent studies have shown very high O_3 concentrations in India, particularly in the Indo-Gangetic Plains (IGP) region, which is one of the most fertile agricultural land areas facing high pollution and population loads [13–15]. By 2030, the population of India is projected to increase further by 300 million people (United Nations Population Division [16,17]. Therefore, due to limited arable land, food security in India is under threat. The selection of O_3-tolerant crops or cultivars can be an important strategy for food security in the area suffering from high-O_3 concentrations [3].

O_3 can adversely affect crop productivity either directly causing the oxidative damage as a result of the production of the reactive oxygen species (ROS) or indirectly as a greenhouse gas [3]. O_3 enters the leaves mainly through stomatal pores. After entry, O_3 is rapidly dissolved in the apoplast and it generates ROS that finally causes an imbalance in the redox status of the cells. This eventually leads to cellular damage or programmed cell death [18–20]. O_3 modifies the plant metabolism by adversely affecting the photosynthetic carbon assimilation, stomatal regulation and plant growth leading to reduced crop yield [21,22].

India is a major producer of wheat (*Triticum aestivum*) accounting 12% of the total wheat production in the world [23]. Wheat is sensitive to O_3-induced damage [17,24]. The critical O_3 level for 5% yield reduction, a three-month Accumulated Ozone exposure over the Threshold of 40 ppb (AOT40) for three months is 3 ppm h, for wheat [25]. Recently, Ghude et al. [24] estimated the production loss of 3.5 ± (0.8) million tons of wheat in India. Despite the obvious sensitivity of wheat to O_3, large scale screening of Indian wheat cultivars in the field conditions has been scarcely done due to technical limitations [26,27]. So far, responses of 14 wheat cultivars cultivated in India have been tested in open top chambers for their O_3 sensitivity to elevated O_3 of 30 ppb with respect to ambient O_3, resulting in lower gas exchange rates, biomass, and yield [27].

A chemical protectant, EDU [ethylenediurea; N-(2-2-oxo-1-imidazolidinyl) ethyl]-N′- phenyl urea N-(2-2-oxo-1-imidazolidinyl) ethyl]-N′-phenyl urea) was first introduced by Carnahan et al. [28]. Thereafter, several reports have shown that the use of EDU can specifically prevent the O_3 injury as well as decrease the growth and yield losses in the ambient field conditions [13,15,29–31]. EDU application is a useful and cost-effective method for the large-scale screening of plant materials for ozone-tolerance/sensitivity, particularly in remote areas, lacking electricity and infrastructures for O_3 exposure or its removal [32].

In this study, we compared the productivity and performance of 11 Indian wheat cultivars throughout the growing season in the ambient field conditions at two high-ozone experimental sites, NBRI (urban) and Banthra (semi-urban), in the IGP region of India. EDU application was used as a research tool to protect the plants against high-ozone stress at both experimental sites. The plants with and without the application of EDU were measured for growth, gas-exchange antioxidants and yield attributes in two developmental phases and two sites. The aim was (1) to classify the cultivars into four adaptation groups, according to their yield (grain weight), and thereafter (2) to indicate the key parameters linked to adaptation strategies for each adaptation group.

2. Materials and Methods

2.1. Experimental Sites and Plant Material

A field study was carried out from 6 December 2011 to 29 March 2012 at two experimental sites (1) the Botanical garden of the CSIR-National Botanical Research Institute, NBRI (26°55′ N, 80°59′ E) in Lucknow and (2) at the Banthra (26°45′ N, 80°53′ E) approximately 25 km from the Lucknow city, India (Figure 1). The NBRI site represents an urban site with soil type of sandy loam (sand 50%, silt 33%, clay 17%; pH 8.4 and electrical conductivity 231.1 μS cm^{-1}), and Banthra a semi-urban site with silt clay loam (sand 14.5%, silt 53.5%, clay 32%; pH 8.4 and electrical conductivity 219 μS cm^{-1}) soil. Eleven locally important wheat (*Triticum aestivum* L.) cultivars were chosen for the present study (Table S1).

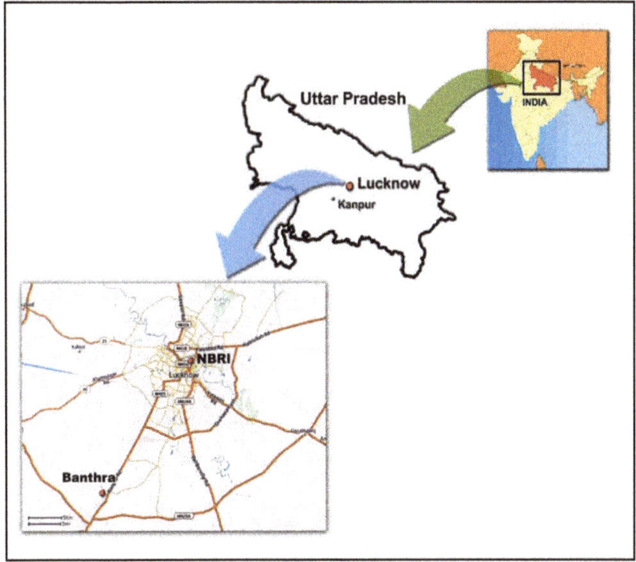

Figure 1. The location of experimental sites in Lucknow, state of Uttar Pradesh, India.

2.2. Experimental Design and EDU Application

The field size was 400 m^2 at the both experimental sites. The field was divided into six plots; three for ambient O_3-grown (non-EDU treated) and three for EDU-treated plants, where non-EDU treated were sprayed with water. Each plot had 11 subplots of 1.5 × 1.5 m in dimension, one for each of the 11 cultivars. The distance between the subplots was 0.5 m. Seeds were sown in each subplot in rows with 25 cm spacing. The recommended dose of NPK fertilisation (120:60:60 kg ha^{-1}) was applied during the field preparation. At first, nitrogen was applied as basal dose which included a full dose of potassium and phosphorus, the second and third doses of nitrogen were applied after 30 and 60 days of sowing (DAS) as top dressing.

EDU was applied at 300 ppm concentration as a foliar spray to individual plants until its entire foliage was visibly saturated. EDU treatment was started after 15 DAS and continued at an interval of 15 days until the final harvest phase. Application of EDU was done on a cloud free day to avoid risk of washing away. The choice of 300 ppm EDU concentration was based on the earlier experiments by Feng et al. [33] suggesting the concentrations at 200–400 ppm range as the most effective in ameliorating effects of high-O_3 concentration in field conditions. Paoletti et al. [34] also demonstrated that 300 ppm of EDU concentration was effective to halt the O_3-induced ROS formation in *Phaseolus vulgaris*. EDU was obtained from Prof. W.J. Manning, University of Massachusetts, Amherst, USA.

2.3. Ozone Monitoring

Ambient O_3 monitoring was carried out with a 2B Tech Ozone Monitor (106-L) for 8 h day^{-1} (9.00 to 17.00) regularly at the NBRI site, and on weekly basis at the Banthra site using the same device (Figure 2). For the NBRI site, the AOT40 (accumulated exposure over a threshold of 40 ppb) exposure index for the O_3 concentration was calculated as described by De Leeuw and Zantvoort [35].

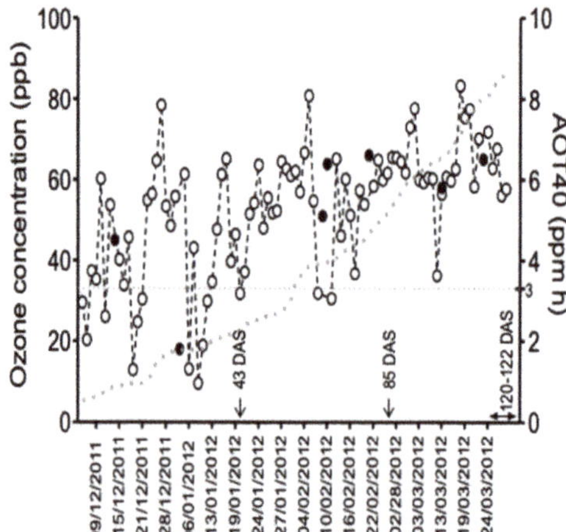

Figure 2. Variation in 8 h (9:00 to 17:00) average ozone concentration at NBRI (open circles) and Banthra (closed circles) sites. Accumulated Ozone exposure over the Threshold of 40 ppb (AOT40) accumulation during the experimental period (6 December 2011 to 29 March 2012) is indicated by grey dashed line. Horizontal line indicates the AOT40 threshold for wheat (3 ppm h). Arrows denote the sampling dates for analyses at vegetative phase (43 days of sowing (DAS)), flowering phase (85 DAS) and harvest phase (120–122 DAS).

2.4. Biomass and Yield Attributes

Harvest index (HI, the ratio of grain yield and the above ground biomass at maturity), 1000 grain weight, grain weight plant^{-1}, and inflorescence weight plant^{-1}, were measured from three randomly selected plants for each cultivar in both treatments (n = 3) at 120–121 DAS at NBRI and 122 DAS at the Banthra site. Above ground biomass was measured for three plants for each cultivar in both treatments at the harvest phase (n = 3).

2.5. Physiological and Biochemical Measurements

The second youngest fully mature leaves were measured for photosynthetic rate (A), stomatal conductance (g_s), and the maximum quantum yield of primary PSII photochemistry (F_v/F_m, the ratio of variable fluorescence to maximum chlorophyll fluorescence) with Li-COR 6400, gas exchange portable photosynthesis system (Li-COR, Lincoln, Nebraska, USA) with a fluorescence chamber (LFC6400-40; Li-COR). Three randomly selected plants of each wheat cultivar in each treatment were measured (n = 3) at the vegetative phase (42–45 DAS) and at the flowering phase at (84–87 DAS) at both experimental sites. The CO$_2$ level inside the leaf cuvette was maintained as 400 μmol mol^{-1}, photosynthetic photon flux density was 1200 μmol mol^{-1}, and leaf temperature was 25 °C.

Leaf samples were collected for the biochemical analyses twice: at the vegetative phase at 43 DAS and at the flowering phase at 85 DAS. The measurements were performed on three randomly selected plants within each cultivar for each treatment (n = 3). Leaf samples were frozen in liquid nitrogen and stored at −80 °C until further analyses. Total chlorophyll content was calculated using equation given in Arnon [36]. The total carotenoid content was calculated from the absorbance values at 480 and 510 nm according to Parsons et al. [37].

Lipid peroxidation in the leaf tissue was determined as 2-thiobarbituric acid (TAB) reactive metabolite, mainly malondialdehyde, Heath and Packer [38]. The Bradford [39] method was used to measure the protein concentration using bovine serum albumin (BSA sigma) as the concentration standard. Superoxide dismutase activity (SOD) was measured using the photochemical NBT method, Beyer and Fridovich [40]. Catalase (CAT) activity was measured by following the reduction in the absorbance at 240 nm as H_2O_2 was consumed Rao et al. [41]. Reduced glutathione (GSH) and oxidized glutathione (GSSG) content were measured by enzyme recycling assay as illustrated by Griffith [42].

2.6. Statistical Analyses

To test the effects of EDU treatment, cultivar and their interaction, two-way ANOVA was performed with SPSS software (SPSS Inc., version 21.0), separately for the vegetative, flowering and the harvest phase and two experimental sites.

To test the differences in the grain weight plant^{-1} for all the 11 tested cultivars, a linear regression was conducted between the mean grain weight plant^{-1} of all the cultivars at each experimental site and EDU treatment (as a numerical measure of the overall quality of the environment) and the individual grain weight of each of the 11 cultivars in the experimental site and treatment combinations. This technique was originally used by Finlay and Wilkinson [43], in order to test the performance of barley in different environments and time scale.

The cultivars were classified in four groups (a–d) based on whether the mean grain weight plant^{-1} of each cultivar was above or below the mean grain weight plant^{-1} of all cultivars (site-mean) in the two environments (NBRI and Banthra). The site mean had a regression coefficient of 1 and the cultivars with clearly higher or lower regression coefficient were considered sensitive or insensitive to environmental change, respectively. The groups were named as: (a) "Well-adapted cultivars" that had an above average grain weight plant^{-1} in all environments. (b) "Poorly adapted cultivars" that had a below average grain weight plant^{-1} in all environments (c) "Cultivars adapted to high yield environments" whose grain weight plant^{-1} was higher than the mean in high-yield environments, but lower than the mean in low-yield environments: They had a regression coefficient clearly higher than 1 indicating strong environmental response in yield. (d) "Cultivars adapted to low yield environments" whose grain weight plant^{-1} was higher than the mean in low yield environments, but lower than mean in high-yield environments. Their regression coefficient was less than 1. Details of the technique used have been illustrated in Pandey et al. [32]. The Spearman correlation of grain yield plant^{-1} with the other measured parameters was tested. The analysis was performed separately for pooled data with all the cultivars and for each cultivar response group.

3. Results

Detailed O_3 data was collected at the NBRI site throughout the study. Less frequent measurements from the Banthra site followed the same pattern. Daily mean O_3 concentrations were above 40 ppb during most of the growing season for wheat, especially during the flowering phase (in February and March), although high daily O_3 concentrations were observed throughout the experiment (Figure 2). The average O_3 concentrations (day time average based on hourly values between 09:00 and 17:00 h) of 45, 45, 57 and 65 ppb were recorded for December, January, February and March, respectively. The average ambient O_3 concentration was 52.8 ppb and ranged between 9.6 and 83.3 ppb during the growth period of wheat. Accumulated Ozone exposure over the Threshold of 40 ppb (AOT40 exposure) was 8.66 ppm h at NBRI site (Figure 2).

3.1. Yield and Biomass in Response to EDU Treatment

The yield attributes (HI, grain weight plant^{-1} and weight of inflorescence) were generally higher in EDU-treated than in non-EDU treated plants, particularly at NBRI site (Figure 3, Figure S1). However, large variation between cultivars in the response to EDU for all yield parameters was evident by the significant Cv × EDU treatment interaction in ANOVA (Table 1 and Figure S1).

Biomass was significantly higher with EDU treatment than in non-EDU treated plants for all the cultivars at the NBRI site (Figure 3A). Since both the grain yield and the biomass were higher in response to EDU, HI was slightly lower in response to EDU treatment at NBRI site (Figure 3A). At Banthra, biomass decreased, and grain yield remained the same, and thus HI improved with EDU treatment, which is indicated by the median in the box-plot suggesting that more than 50% of the cultivars showed improved HI (Figure 3B).

The 11 cultivars represented different response groups in a regression analysis of grain weight plant^{-1}: (a) three cultivars (Kundan, WR544 and PBW550) were generally well-adapted with above-average yield (Figure 4A), (b) three cultivars (PBW373, PBW154, HUW234) were poorly-adapted with below-average yield (Figure 4A); (c) two cultivars (PBW343, LOK1) were adapted to low-yield environments (Figure 4B); and (d) three cultivars (PBW502, WH711 and DBW17) were adapted to high-yield environments (Figure 4B). The cultivars adapted to high-yield environments (WH711 and DBW17) had the poorest grain yield of the whole experiment at the NBRI site at ambient O_3 conditions.

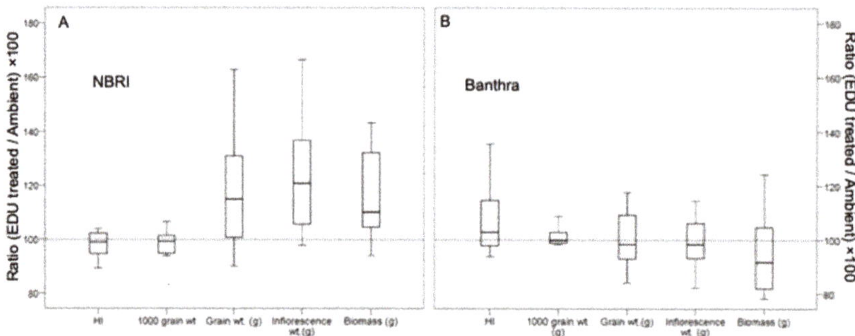

Figure 3. The effect of Ethylenediurea (EDU) treatment on various yield attributes across all the cultivars {(based on relative values, e.g., harvest index with non-EDU/harvest index with EDU treatment) × 100)}. The median of each parameter is shown as the horizontal bar in each box, and the upper and the lower sides of a box represent the first and the third quartile values of the distribution respectively. Harvest index (HI), 1000 grain weight plant^{-1} (1000 grain wt.), grain weight plant^{-1} (grain wt.), inflorescence weight plant^{-1} (wt. of inflorescence) and biomass.

Table 1. F ratios and levels of significance of multivariate ANOVA test for different parameters of all the 11 tested cultivars. Significant results of two-way ANOVA are marked with asterisks (* $p < 0.05$ and ** $p < 0.01$) for Cultivar, Treatment (EDU) and Cultivar × treatment (EDU) at experimental sites; NBRI and Banthra at vegetative, flowering and harvest phase. Superoxide dismutase activity (SOD), catalase activity (CAT), reduced glutathione (GSH), oxidised glutathione (GSSG), malondialdehyde content (MDA), total chlorophyll (Tchl), carotenoid (Caro), photosynthesis (A), stomatal conductance (gs), ratio of variable to maximal chlorophyll fluorescence (Fv/Fm), harvest index (HI), 1000 grain weight plant^{-1} (1000_grain wt), inflorescence weight plant^{-1} (Inflorescence wt) and biomass.

	NBRI						Banthra					
	Vegetative			Flowering			Vegetative			Flowering		
	Cultivar	Treatment (EDU)	Cultivar × treatment (EDU)	Cultivar	Treatment (EDU)	Cultivar × treatment (EDU)	Cultivar	Treatment (EDU)	Cultivar × treatment (EDU)	Cultivar	Treatment (EDU)	Cultivar × treatment (EDU)
SOD	23.36 **	175.10 **	25.63 **	1.86	0.03	1.81 **	48.16 **	956.17 **	28.25 **	45.78 **	0.04	35.04 **
CAT	27.26 **	46.27 **	25.94 **	17.56 **	24.40 **	14.79 **	67.83 **	0.34	17.77 **	169.41 **	307.79 **	108.07 **
GSH	4.71 **	1.82	6.67 **	14.86 **	16.82 **	19.63 **	10.78 **	699.47 **	12.57 **	11.36 **	15.86 **	18.38 **
GSSG	10.80 **	25.17 **	19.12 **	7.62 **	76.36 **	14.48 **	26.24 **	82.41 **	7.02 **	36.15 **	24.56 **	21.41 **
MDA	12.97 **	3.56	8.45 **	46.62 **	12.06 **	14.02 **	17.69 **	238.44 **	26.32 **	18.05 **	82.85 **	5.71 **
T Chl	41.92 **	58.73 **	26.84 **	10.60 **	8.53 **	15.20 **	1614.38 **	2919.22 **	2101.01 **	327.17 **	3314.84 **	128.99 **
Caro	25.94 **	27.69 **	15.56 **	24.52 **	2.47	12.26 **	1549.28 **	1622.39 **	2027.76 **	514.43 **	1993.56 **	158.96 **
A	2.08 *	1.15	4.63 **	17.43 **	18.21 **	23.80 **	11.19 **	33.61 **	11.16 **	25.52 **	2.38	9.98 **
gs	3.22 **	0.37	1.65	13.89 **	3.03	11.31	6.33 **	31.85 **	5.93 **	39.34 **	7.13 **	3.30 **
FvFm	1.79	0.08	1.09	2.97 **	2.98	1.69	0.65	4.10 *	1.51	0.96	0.23	0.93
Harvest parameters							**Harvest parameters**					
HI	16.04 **	6.03 **	2.74 **				67.73 **	22.33 **	15.31 **			
1000_grain wt	401.47 **	55.75 **	51.52 **				331.58 **	62.61 **	31.57 **			
Inflorescence wt	12.58 **	96.61 **	6.80 **				12.56 **	0.66	3.44 **			
Grain_wt	10.01 **	34.61 **	3.93 **				10.78 **	0.00	3.39 **			
Biomass	14.65 **	81.30 **	8.24 **				12.95 **	20.95 **	12.12 **			

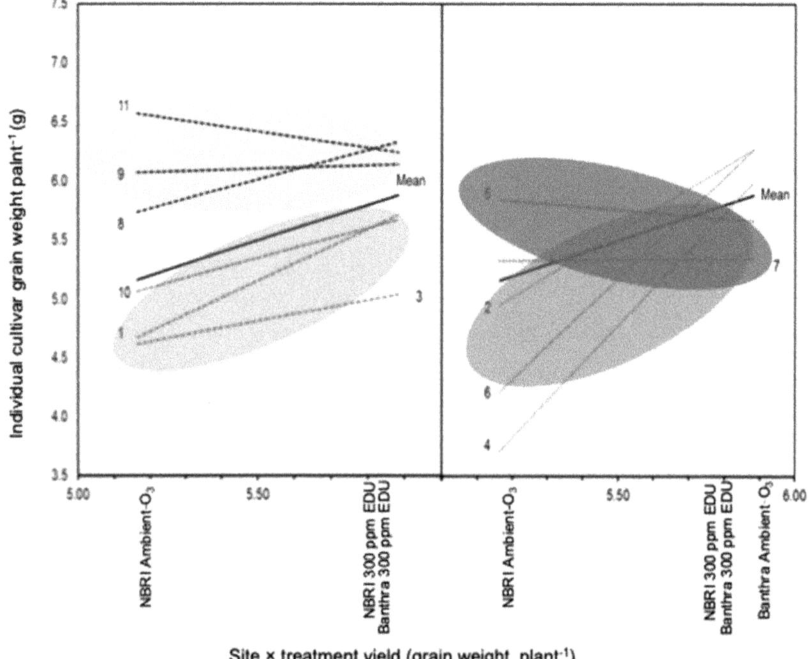

Figure 4. Regressions between the mean grain weight plant^{-1} of all the cultivars at each site and treatment (x-axis) and the individual grain weight plant^{-1} (y-axis) of each of the 11 wheat cultivars. (**A**) Grouping of well adapted (8, 9 and 11) cultivars and poorly adapted (1, 3 and 10) cultivars. (**B**) Grouping of cultivars adapted to low-yield conditions (5 and 7) and cultivars adapted to high-yield (2, 6 and 4) conditions. (1) PBW-373, (2) PBW-502, (3) PBW-154, (4) WH711, (5) PBW-343, (6) DBW-17, (7) LOK-1, (8) KUNDAN, (9) WR-544, (10) HUW-234, (11) PBW-550.

3.2. Gas Exchange and Pigments

Gas exchange was affected by cultivar, EDU, developmental stage and study site. Although impact of EDU on A and g_s was variable among the cultivars as shown by the significant Cv × EDU treatment interactions (Table 1 and Figure S2) EDU-treated plants tended to have higher A and g_s than non-EDU ones (significantly only in Banthra at vegetative phase) (Supplementary Figures S3 and S5).

Cultivars differed from each other also in the contents of pigments (chlorophyll, carotenoids) throughout the experiment at both experimental sites (Figure S4). Contents of chlorophyll and carotenoids differed among the cultivars and treatments in a similar way. Significantly lower contents of chlorophyll and carotenoids were detected in EDU-treated plants than non-EDU ones in the flowering phase at both experimental sites (Figure S5). EDU-treated plants had higher chlorophyll and carotenoid content than the non-EDU treated plants at the vegetative phase at NBRI (Figure S5), whereas they had similar or even lower contents of pigments at the flowering phase at both experimental sites (Figure S5).

3.3. Biochemical Measurements

EDU treatment had a significant effect on all measured biochemical parameters (MDA, CAT, GSH, GSSG, SOD), but the responses varied among the cultivars, developmental phases and experimental sites throughout the experiment (Table 1 and Figures S3, S6 and S7). Lipid peroxidation (MDA content) tended to be lower in EDU-treated plants than non-EDU ones at both experimental sites (except for

Banthra at the vegetative phase) (Figure S7). EDU-treated plants had higher SOD activity and GSH content than non-EDU treated ones in Banthra at vegetative stage, while EDU-treated plants had lower SOD activity than non-EDU treated ones (NBRI, vegetative phase), GSSG content (NBRI, flowering phase; Banthra, vegetative phase) than non-EDU ones (Figure S7).

3.4. Correlations of Measured Parameters and Grain Yield

The strongest positive correlations to grain yield across all response groups were found for inflorescence weight^{-1} and biomass, except for the cultivars adapted to low-yield environments (Table 2). HI showed a positive correlation with grain yield in the cultivars adapted to high-yield environments. The strongest negative correlations to grain yield were found for CAT activity (except for the cultivars adapted to high-yield conditions) and GSSG (except for the cultivars adapted to low-yield conditions) at the flowering phase. The grain yield of the well-adapted cultivars showed a positive correlation with A at the flowering phase (Table 2). The grain yield of the poorly-adapted cultivars showed positive correlation with A and gs at the vegetative phase, followed by strong negative correlations with the CAT, GSSG and MDA at the flowering stage (Table 2). The grain yield of the cultivars adapted to low-yield conditions showed negative correlations with chlorophyll content at the vegetative phase and CAT at the flowering phase (Table 2). In the cultivars adapted to high-yield environments, strong negative correlation was found with SOD at the vegetative phase, accompanied by positive correlations with CAT, GSH and GSSG. At the flowering phase, positive correlation with SOD and contents of chlorophyll and carotenoids were accompanied with negative correlations with CAT, GSSG and gas exchange parameters.

Table 2. Correlation of the different parameters with the grain yield plant^{-1} for the groups assigned from the Finlay method. Significant correlations are in bold. Positive correlations (light grey) and negative (grey) are presented in the table. Superoxide dismutase activity (SOD), catalase activity (CAT), reduced glutathione (GSH), oxidised glutathione (GSSG), malondialdehyde content (MDA), chlorophyll (chl), carotenoid (Caro), photosynthesis (A), stomatal conductance (g_s), ratio of variable to maximal chlorophyll fluorescence (F_v/F_m), harvest index (HI), 1000 grain weight plant^{-1} (1000_grain wt), inflorescence weight plant (Inflorescence wt) and biomass.

Parameters	Cultivars 8,9,11 (Well-adapted)	Cultivars 1,3,10 (Poorly Adapted)	Cultivars 5,7 (Low-yield Condition)	Cultivars 2,4,6 (High-yield Condition)	All Cultivars
Vegetative					
SOD	−0.151	0.011	0.166	**−0.614 **	**−0.313 ***
CAT	0.005	0.294	0.244	0.156	0.311 *
GSH	0.125	0.060	−0.240	−0.149	0.033
GSSG	0.277	**−0.542 **	0.192	**0.490 **	0.325 *
MDA	−0.246	0.163	0.087	−0.007	−0.024
Chl	−0.263	0.230	**−0.472 ***	−0.079	−0.087
Car	−0.320	0.294	−0.401	−0.221	−0.12
A	−0.285	**0.441 **	−0.204	0.132	0.144
g_s	−0.281	**0.344 ***	−0.218	0.245	0.26
F_v/F_m	0.049	−0.055	0.116	0.062	0.06
Flowering					
SOD	−0.033	0.155	−0.364	**0.590 **	0.137
CAT	0.222	**−0.472 **	**−0.408 ***	**−0.629 **	**−0.373 ***
GSH	0.266	0.067	0.236	−0.203	0.191
GSSG	−0.132	**−0.379 ***	0.401	**−0.543 **	**−0.322 ***
MDA	0.010	**−0.618 **	−0.222	−0.178	−0.251
Chl	0.202	−0.010	0.132	**0.384 ***	0.236
Car	−0.117	0.116	0.053	**0.410 ***	0.198
A	**0.383 ***	0.147	−0.078	**−0.533 **	−0.141
gs	0.189	−0.093	−0.140	**−0.653 **	−0.257
FvFm	0.284	0.012	−0.125	−0.098	0.107
Final harvest					

Table 2. *Cont.*

Parameters	Cultivars 8,9,11 (Well-adapted)	Cultivars 1,3,10 (Poorly Adapted)	Cultivars 5,7 (Low-yield Condition)	Cultivars 2,4,6 (High-yield Condition)	All Cultivars
HI	0.209	0.064	0.327	0.659 **	0.426 **
1000_grain wt	−0.082	−0.425 **	−0.097	0.243	−0.246
Inflorescence wt	0.614 **	0.599 **	−0.033	0.695 **	0.737 **
Biomass	0.472 **	0.618 **	0.138	0.694 **	0.776 **

4. Discussion

In this study EDU application was used as ozone-protectant to indicate the severity of the O_3-induced damage in wheat production in an agriculturally important region suffering from high pollution in a highly populated area of India. The O_3 concentration increased gradually during the growing period of wheat from December to March, particularly at the grain filling phase (February to March), which has been considered to be the most sensitive stage to O_3 damage, especially for wheat [44]. The critical three-month O_3 level for wheat (3 ppm h) [25] was not reached at the vegetative phase, but it was attained before the flowering phase resulting in O_3 exposure that was double than the estimated damage threshold by the harvest time. Accordingly, our results indicate the strong impact of O_3 in the flowering and harvest stage. AOT40 values and the average O_3 concentrations were in line with the other studies performed in this region of India reviewed by Oksanen et al. [13] and Ainsworth [3], e.g., with mustard (*Brassica campestris*) [45] and clover (*Trifolium alexandrium* L.) [27].

4.1. Biomass, Allocation Strategies and Grain Yield

Our experiment showed clear differences in antioxidant and gas exchange parameters among the cultivars, adaptation groups and the two developmental phases. These results can be linked to O_3-tolerance and O_3-defence strategies, because plants treated with EDU application can be regarded to represent clean-air controls. O_3 tolerance of the plants can be linked to two important strategies, the regulation of stomatal conductance and the potential to detoxify the ROS generated in the course O_3 degradation [14,46–48]. Previous studies have also indicated that O_3-sensitive cultivars tend to allocate more of their resources to defense actions in response to O_3 limiting biomass [27,45,49,50].

In the present study, biomass accumulation showed a positive significant correlation with the grain yield. The associations between the grain yield and other parameters in this study indicated that the grain yield of the well-adapted cultivars was not associated with the biochemical parameters, but rather the higher the yield was correlating with high photosynthesis (*A*) at the flowering stage (Table 2). Poorly-adapted cultivars showed positive correlations with gas exchange rates during the vegetative stage, which may indicate high O_3 uptake, accompanied by weak antioxidative defense through GSSG and CAT. Cultivars adapted to low-yield conditions were limited by chlorophyll content and poor defense by CAT. Cultivars adapted to high-yield conditions (including EDU protection) are relying on high antioxidative defense through CAT, GSH, GSSG during the vegetative stage, with negative correlations (trade-off) with SOD. At the flowering stage, antioxidant status was reversed and accompanied by low gas exchange rates but high contents of chlorophyll and carotenoids. Thus, our study indicated that defense strategies are complex and may vary during the development. Clearly, low grain yield in our experiment was associated with low CAT activity but high GSSG content at the flowering phase for most of the cultivar groups. GSSG content has been shown to accumulate in response to O_3, as well as GSH content and total glutathione [51]. Higher total glutathione content has been associated with high tolerance to O_3 in poplar trees [51]. Singh et al. [49] have exposed 14 wheat cultivars to elevated (ambient +30 ppb) O_3 and classified them in three different classes: sensitive, intermediately sensitive, and tolerant cultivars based on the cumulative stress response matrix using growth, physiological and yield. Two cultivars included also in our study, i.e., the well-adapted Kundan and the high-yield environment adapted PBW502, were classified by Singh et al. [49] as O_3-tolerant and intermediately sensitive, respectively, which was in accordance with our classification despite the different grouping method. Reduced biomass due to O_3 stress may also be attributed to several other physiological and biochemical events in the developmental phase of the plants, for example decline in Rubisco activity [50]. Pleijel and Uddling [52] reported that O_3 can significantly reduce the proportion of above-ground biomass converted to grains, on the contrary, in the present study, biomass accumulation showed a positive significant correlation with the grain yield.

The higher biomass and yield with EDU treatment compared to ambient field conditions reflect the positive effects of EDU in those parameters, which are often negatively affected by O_3 [53,54]. In a meta-analysis by Feng et al. [33] the increase of the above-ground biomass by 6.7% was reported with EDU treatment. Similar biomass enhancements with EDU treatment under high O_3 have been reported in wheat (*Triticum aestivum* L.) [55], mustard (*Brassica campestris* L.) [45], rice (*Oryza sativa* L.) [32] and pea (*Phaseolus vulgaris* L.) [56]. In addition to positive impact of EDU, in the present study indicated that the resource allocation strategies in response with EDU differed among the wheat cultivars and between experimental sites. At NBRI, the wheat plants showed more efficient resource allocation towards grains in response to EDU treatment which was accompanied by improved biomass and slight decrease in HI. However, at Banthra, the biomass was lower with EDU-treated plants than non-EDU ones and HI was slightly improved (due to decrease in above-ground biomass) and grain yield was not higher with EDU-treatment than in non-EDU treated plants.

4.2. EDU as a Tool to Reveal Ozone Impact

In the present study, EDU responses were not only limited to growth, gas-exchange or the biochemical parameters, but also showed that the prevailing O_3 concentration had an adverse effect on yield attributes, reflected as reduced grain yield at the harvest phase. EDU-mediated increase in the antioxidant defense (SOD, CAT, APX and GR), growth parameters, biomass, and yield attributes have been reported in previous studies under high O_3 conditions. The activation of the antioxidative defense and EDU responses, however, are related to severity of the oxidative stress [11,13,29,30,32,45,50,55,57–60].

The positive impact of EDU on gas exchange and photosynthesis related parameters is not well-established, as evidenced also in our experiment showing the high variation among the cultivars. Feng et al. [33], Hassan et al. [61] and Manning et al. [57] have reported that EDU did not show any clear effect on the g_s and A. On the other hand, a positive impact of EDU on g_s in rice [59] and wheat [26] and on A, g_s, light reaction and F_v/F_m in pea (*Phaseolus vulgaris* L.) [56] have been reported, especially on O_3 sensitive cultivars, which is in accordance with our results at vegetative phase at Banthra. The higher chlorophyll content in non-EDU grown plants than in EDU-treated plants during the flowering phase (NBRI and Banthra) may indicate O_3-induced compensatory responses in the newly formed leaves, as all the measurements were conducted on the youngest fully mature leaves. Such compensatory responses appearing as increased shoot weight plant^{-1} [32,62] leaf greenness, and photosynthetic adjustment [32,63] have been reported in response to high O_3. The similarity of responses between chlorophyll and carotenoid content across the cultivars was expected because of the similarity in the regulation of their biosynthesis [64].

Our experiment demonstrates the applicability of EDU as a surface treatment in large-scale screening for O_3-tolerance in wheat cultivars in different environments. A recent study by Ashrafuzzaman et al. [31] also suggest that EDU did not interfere with the gene-regulations and did not affect the tolerance of the plants to other abiotic stresses, such as iron toxicity, zinc deficiency and salinity stresses, under O_3-stress conditions in rice, which further strengthen the potential use of EDU in field conditions. Several other studies with rice [32,55,59] and wheat [50,55] have also reported the usefulness of EDU in the field conditions in identifying O_3-tolerant cultivars. In the present study, the EDU-responses varied not only among the cultivars, but also due to growth phase and experimental site, as reported also in pea [65], mustard [45], rice [32,59] and wheat [50]. Although the exact mechanism for the mode of action of EDU still unclear, it has been demonstrated that the nitrogen present in EDU has no role in fertilization, growth regulation, or grain yield under O_3-free conditions [2,59].

EDU is currently not commercially available and can thus be applied for research purposes only. Earlier studies with EDU suggest the range between 100 and 300 ppm (100 to 300 mg L^{-1}) to be the most effective concentration in ameliorating negative effects against O_3 without having any toxic effects of its own [2]. The concentration of 300 mg EDU L^{-1} was also recommended as the upper limit for toxicity in a toxicological bioassay in *Lemna minor* [66]. Manning et al. [30] reported that EDU did not show any constitutive effects on the crops in O_3-free control conditions.

5. Conclusions

Our experiment with EDU application at two different high-ozone environments indicated high variation in the resource allocation and the defense strategies in the Indian wheat cultivars. The well-adapted cultivars in our study, i.e., Kundan, WR544 and PBW550 showed a high yield regardless of the site in the IGP area of India. In these well-adapted cultivars, the grain yield was related to high net assimilation (*A*) at the flowering stage of the development and high biomass accumulation at the end of the experiment. On the other hand, all other response groups showed high stomatal conductance and net assimilation at vegetative phase and low antioxidant defense (CAT activity, glutathione content) at vegetative and flowering phases. The cultivars that were able to maintain high antioxidative defense and net assimilation capacity ended up with higher yield indicating higher ozone tolerance. It is clear that a wide screening of wheat cultivars is necessary to improve food-security for crops in areas experiencing high O_3 concentrations. Based on our results, high throughput screening will reveal high differences among cultivars and help to find the key parameters to be studied.

Supplementary Materials: The following are available online at http://www.mdpi.com/2225-1154/7/2/23/s1.

Author Contributions: E.O. conceived the project. A.K.P. and B.M. performed the experimental part. V.P. supervised the experiments. A.K.P., S.K.-S. (Sarita Keski-Saari), S.K.-S. (Sari Kontunen-Soppela), E.O., V.P. performed writing reviewing and editing.

Funding: This study was funded by the Academy of Finland, project no 138161. AKP also acknowledges FinCEAL Plus Research Visit Grant project No. 51770 for the financial support to travel to Finland.

Acknowledgments: The authors express a sincere thanks to W.J. Manning, Department of Plant, Soil and Insect Science, University of Massachusetts, Amherest, USA, for providing EDU for this study. AP and BM thank the Academy of Finland and for the financial support. Jenna Lihavainen is acknowledged for long insightful discussions.

Conflicts of Interest: The authors declare no conflict of interest.

References

1. Agathokleous, E.; Saitanis, C.J.; Koike, T. Tropospheric O_3, the nightmare of wild plants: A review study. *J. Agric. Meteorol.* **2015**, *71*, 142–152. [CrossRef]
2. Agathokleous, E. Perspectives for elucidating the ethylenediurea (EDU) mode of action for protection against O_3 phytotoxicity. *Ecotoxicol. Environ. Saf.* **2017**, *142*, 530–537. [CrossRef] [PubMed]
3. Ainsworth, E.A. Understanding and improving global crop response to ozone pollution. *Plant J.* **2017**, *90*, 886–897. [CrossRef]
4. Emberson, L.D.; Büker, P.; Ashmore, M.R.; Mills, G.; Jackson, L.S.; Agrawal, M.; Atikuzzaman, M.D.; Cinderby, S.; Engardt, M.; Jamir, C.; et al. A comparison of North American and Asian exposure–response data for ozone effects on crop yields. *Atmos. Environ.* **2009**, *43*, 1945–1953. [CrossRef]
5. Tai, A.P.; Martin, M.V.; Heald, C.L. Threat to future global food security from climate change and ozone air pollution. *Nat. Clim. Chang.* **2014**, *4*, 817. [CrossRef]
6. Fuhrer, J. Ozone risk for crops and pastures in present and future climates. *Naturwissenschaften* **2009**, *96*, 173–194. [CrossRef] [PubMed]
7. Tai, A.P.K.; Val Martin, M. Impacts of ozone air pollution and temperature extremes on crop yields: Spatial variability, adaptation and implications for future food security. *Atmos. Environ.* **2017**, *169*, 11–21. [CrossRef]
8. Avnery, S.; Mauzerall, D.L.; Fiore, A.M. Increasing global agricultural production by reducing ozone damages via methane emission controls and ozone-resistant cultivar selection. *Glob. Chang. Biol.* **2013**, *19*, 1285–1299. [CrossRef]

9. Feng, Z.; Hu, E.; Wang, X.; Jiang, L.; Liu, X. Ground-level O_3 pollution and its impacts on food crops in China: A review. *Environ. Pollut.* **2015**, *199*, 42–48. [CrossRef]

10. Tiwari, S.; Rai, R.; Agrawal, M. Annual and seasonal variations in tropospheric ozone concentrations around Varanasi. *Int. J. Remote Sens.* **2008**, *29*, 4499–4514. [CrossRef]

11. Tiwari, S. Ethylenediurea as a potential tool in evaluating ozone phytotoxicity: A review study on physiological, biochemical and morphological responses of plants. *Environ. Sci. Pollut. Res.* **2017**, *24*, 14019–14039. [CrossRef] [PubMed]

12. Brauer, M. The global burden of disease from air pollution. In Proceedings of the 2016 AAAS Annual Meeting, Washington, DC, USA, 11–15 February 2016.

13. Oksanen, E.; Pandey, V.; Pandey, A.K.; Keski-Saari, S.; Kontunen-Soppela, S.; Sharma, C. Impacts of increasing ozone on Indian plants. *Environ. Pollut.* **2013**, *177*, 189–200. [CrossRef] [PubMed]

14. Pandey, A.K.; Ghosh, A.; Agrawal, M.; Agrawal, S.B. Effect of elevated ozone and varying levels of soil nitrogen in two wheat (*Triticum aestivum* L.) cultivars: Growth, gas-exchange, antioxidant status, grain yield and quality. *Ecotoxicol. Environ. Saf.* **2018**, *158*, 59–68. [CrossRef] [PubMed]

15. Tiwari, S.; Agrawal, M. Effect of Ozone on Physiological and Biochemical Processes of Plants. In *Tropospheric Ozone and its Impacts on Crop Plants*; Springer: Cham, Switzerland, 2018; pp. 65–113.

16. United Nations Population Division (2010) World Population Prospects, the 2010 Revision. Available online: http://esa.un.org/unpd/wpp/ (accessed on 20 April 2018).

17. Avnery, S.; Mauzerall, D.L.; Liu, J.; Horowitz, L.W. Global crop yield reductions due to surface ozone exposure: 2. Year 2030 potential crop production losses and economic damage under two scenarios of O_3 pollution. *Atmos. Environ.* **2011**, *45*, 2297–2309. [CrossRef]

18. Castagna, A.; Ranieri, A. Detoxification and repair process of ozone injury: From O_3 uptake to gene expression adjustment. *Environ. Pollut.* **2009**, *157*, 1461–1469. [CrossRef]

19. Fiscus, E.L.; Booker, F.L.; Burkey, K.O. Crop responses to ozone: Uptake, modes of action, carbon assimilation and partitioning. *Plant Cell Environ.* **2005**, *28*, 997–1011. [CrossRef]

20. Vahisalu, T.; Puzõrjova, I.; Brosché, M.; Valk, E.; Lepiku, M.; Moldau, H.; Pechter, P.; Wang, Y.S.; Lindgren, O.; Salojärvi, J.; Loog, M. Ozone-triggered rapid stomatal response involves the production of reactive oxygen species, and is controlled by SLAC1 and OST1. *Plant J.* **2010**, *62*, 442–453. [CrossRef]

21. Black, V.J.; Black, C.R.; Roberts, J.A.; Stewart, C.A. Tansley Review No. 115 Impact of ozone on the reproductive development of plants. *New Phytol.* **2000**, *147*, 421–447. [CrossRef]

22. Feng, Z.; Kobayashi, K. Assessing the impacts of current and future concentrations of surface ozone on crop yield with meta-analysis. *Atmos. Environ.* **2009**, *43*, 1510–1519. [CrossRef]

23. Tomer, R.; Bhatia, A.; Kumar, V.; Kumar, A.; Singh, R.; Singh, B.; Singh, S.D. Impact of elevated ozone on growth, yield and nutritional quality of two wheat species in Northern India. *Aerosol Air Qual. Res.* **2015**, *15*, 329–340. [CrossRef]

24. Ghude, S.D.; Jena, C.; Chate, D.M.; Beig, G.; Pfister, G.G.; Kumar, R.; Ramanathan, V. Reductions in India's crop yield due to ozone. *Geophys. Res. Lett.* **2014**, *41*, 5685–5691. [CrossRef]

25. Mills, G.; Buse, A.; Gimeno, B.; Bermejo, V.; Holland, M.; Emberson, L.; Pleijel, H. A synthesis of AOT40-based response functions and critical levels of ozone for agricultural and horticultural crops. *Atmos. Environ.* **2007**, *41*, 2630–2643. [CrossRef]

26. Singh, S.; Agrawal, S.B.; Agrawal, M. Differential protection of ethylenediurea (EDU) against ambient ozone for five cultivars of tropical wheat. *Environ. Pollut.* **2009**, *157*, 2359–2367. [CrossRef] [PubMed]

27. Singh, S.; Singh, P.; Agrawal, S.B.; Agrawal, M. Use of Ethylenediurea (EDU) in identifying indicator cultivars of Indian clover against ambient ozone. *Ecotoxicol. Environ. Saf.* **2018**, *147*, 1046–1055. [CrossRef]

28. Carnahan, J.E.; Jenner, E.L.; Wat, E.K.W. Prevention of ozone injury to plants by a new protectant chemical. *Phytopathology* **1978**, *68*, 1229. [CrossRef]

29. Paoletti, E.; Contran, N.; Manning, W.J.; Ferrara, A.M. Use of the antiozonant ethylenediurea (EDU) in Italy: Verification of the effects of ambient ozone on crop plants and trees and investigation of EDU's mode of action. *Environ. Pollut.* **2009**, *157*, 1453–1460. [CrossRef] [PubMed]

30. Manning, W.J.; Paoletti, E.; Sandermann Jr, H.; Ernst, D. Ethylenediurea (EDU): A research tool for assessment and verification of the effects of ground level ozone on plants under natural conditions. *Environ. Pollut.* **2011**, *159*, 3283–3293. [CrossRef] [PubMed]

31. Ashrafuzzaman, M.; Haque, Z.; Ali, B.; Mathew, B.; Yu, P.; Hochholdinger, F.; de Abreu Neto, J.B.; McGillen, M.R.; Ensikat, H.J.; Manning, W.J.; et al. Ethylenediurea (EDU) mitigates the negative effects of ozone in rice: Insights into its mode of action. *Plant Cell Environ.* **2018**, *41*, 2882–2898. [CrossRef]

32. Pandey, A.K.; Majumder, B.; Keski-Saari, S.; Kontunen-Soppela, S.; Mishra, A.; Sahu, N.; Pandey, V.; Oksanen, E. Searching for common responsive parameters for ozone tolerance in 18 rice cultivars in India: Results from ethylenediurea studies. *Sci. Total Environ.* **2015**, *532*, 230–238. [CrossRef]

33. Feng, Z.; Wang, S.; Szantoi, Z.; Chen, S.; Wang, X. Protection of plants from ambient ozone by applications of ethylenediurea (EDU): A meta-analytic review. *Environ. Pollut.* **2010**, *158*, 3236–3242. [CrossRef]

34. Paoletti, E.; Castagna, A.; Ederli, L.; Pasqualini, S.; Ranieri, A.; Manning, W.J. Gene expression in snapbeans exposed to ozone and protected by ethylenediurea. *Environ. Pollut.* **2014**, *193*, 1–5. [CrossRef] [PubMed]

35. De Leeuw, F.A.A.M.; Van Zantvoort, E.D.G. Mapping of exceedances of ozone critical levels for crops and forest trees in the Netherlands: Preliminary results. *Environ. Pollut.* **1997**, *96*, 89–98. [CrossRef]

36. Arnon, D.I. Copper enzymes in isolated chloroplasts. Polyphenoloxidase in *Beta vulgaris*. *Plant Physiol.* **1949**, *24*, 1–15. [CrossRef]

37. Parsons, T.R.; Maita, Y.; Lalli, C.M. *A Manual of Chemical and Biological Methods for Seawater Analysis*; Pergamon Press: Oxford, UK, 1984.

38. Heath, R.L.; Packer, L. Photoperoxidation in isolated chloroplasts: I. Kinetics and stoichiometry of fatty acid peroxidation. *Arch. Biochem. Biophys.* **1968**, *125*, 189–198. [CrossRef]

39. Bradford, M.M. A rapid and sensitive method for the quantitation of microgram quantities of protein utilizing the principle of protein-dye binding. *Anal. Biochem.* **1976**, *72*, 248–254. [CrossRef]

40. Beyer, W.F., Jr.; Fridovich, I. Assaying for superoxide dismutase activity: Some large consequences of minor changes in conditions. *Anal. Biochem.* **1987**, *161*, 559–566. [CrossRef]

41. Rao, M.V.; Paliyath, G.; Ormrod, D.P. Ultraviolet-B-and ozone-induced biochemical changes in antioxidant enzymes of Arabidopsis thaliana. *Plant Physiol.* **1996**, *110*, 125–136. [CrossRef] [PubMed]

42. Griffith, O.W. Determination of glutathione and glutathione disulfide using glutathione reductase and 2-vinylpyridine. *Anal. Biochem.* **1980**, *106*, 207–212. [CrossRef]

43. Finlay, K.W.; Wilkinson, G.N. The analysis of adaptation in a plant-breeding programme. *Aust. J. Agric. Res.* **1963**, *14*, 742–754. [CrossRef]

44. Picchi, V.; Iriti, M.; Quaroni, S.; Saracchi, M.; Viola, P.; Faoro, F. Climate variations and phenological stages modulate ozone damages in field-grown wheat. A three-year study with eight modern cultivars in Po Valley (Northern Italy). *Agric. Ecosyst. Environ.* **2010**, *135*, 310–317. [CrossRef]

45. Pandey, A.K.; Majumder, B.; Keski-Saari, S.; Kontunen-Soppela, S.; Pandey, V.; Oksanen, E. Differences in responses of two mustard cultivars to ethylenediurea (EDU) at high ambient ozone concentrations in India. *Agric. Ecosyst. Environ.* **2014**, *196*, 158–166. [CrossRef]

46. Brosché, M.; Merilo, E.B.E.; Mayer, F.; Pechter, P.; Puzõrjova, I.; Brader, G.; Kangasjärvi, J.; Kollist, H. Natural variation in ozone sensitivity among Arabidopsis thaliana accessions and its relation to stomatal conductance. *Plant Cell Environ.* **2010**, *33*, 914–925. [CrossRef] [PubMed]

47. Dizengremel, P.; Le Thiec, D.; Bagard, M.; Jolivet, Y. Ozone risk assessment for plants: Central role of metabolism-dependent changes in reducing power. *Environ. Pollut.* **2008**, *156*, 11–15. [CrossRef] [PubMed]

48. Felzer, B.S.; Cronin, T.; Reilly, J.M.; Melillo, J.M.; Wang, X. Impacts of ozone on trees and crops. *C. R. Geosci.* **2007**, *339*, 784–798. [CrossRef]

49. Singh, A.A.; Fatima, A.; Mishra, A.K.; Chaudhary, N.; Mukherjee, A.; Agrawal, M.; Agrawal, S.B. Assessment of ozone toxicity among 14 Indian wheat cultivars under field conditions: Growth and productivity. *Environ. Monit. Assess.* **2018**, *190*, 190. [CrossRef]

50. Gupta, S.K.; Sharma, M.; Majumder, B.; Maurya, V.K.; Lohani, M.; Deeba, F.; Pandey, V. Impact of Ethylene diurea (EDU) on growth, yield and proteome of two winter wheat varieties under high ambient ozone phytotoxicity. *Chemosphere* **2017**, *196*, 161–173. [CrossRef] [PubMed]

51. Dumont, J.; Keski-Saari, S.; Keinänen, M.; Cohen, D.; Ningre, N.; Kontunen-Soppela, S.; Baldet, P.; Gibon, Y.; Dizengremel, P.; Vaultier, M.N.; et al. Ozone affects ascorbate and glutathione biosynthesis as well as amino acid contents in three Euramerican poplar genotypes. *Tree Physiol.* **2014**, *34*, 253–266. [CrossRef]

52. Pleijel, H.; Uddling, J. Yield vs. Quality trade-offs for wheat in response to carbon dioxide and ozone. *Glob. Chang. Biol.* **2012**, *18*, 596–605. [CrossRef]

53. Mishra, A.K.; Rai, R.; Agrawal, S.B. Individual and interactive effects of elevated carbon dioxide and ozone on tropical wheat (*Triticum aestivum* L.) cultivars with special emphasis on ROS generation and activation of antioxidant defence system. *Indian J. Biochem. Biophys.* **2013**, *50*, 139–149.

54. Saitanis, C.J.; Bari, S.M.; Burkey, K.O.; Stamatelopoulos, D.; Agathokleous, E. Screening of Bangladeshi winter wheat (*Triticum aestivum* L.) cultivars for sensitivity to ozone. *Environ. Sci. Pollut. Res.* **2014**, *21*, 13560–13571. [CrossRef]

55. Wang, X.; Zheng, Q.; Yao, F.; Chen, Z.; Feng, Z.; Manning, W.J. Assessing the impact of ambient ozone on growth and yield of a rice (*Oryza sativa* L.) and a wheat (*Triticum aestivum* L.) cultivar grown in the Yangtze Delta, China, using three rates of application of ethylenediurea (EDU). *Environ. Pollut.* **2007**, *148*, 390–395. [CrossRef] [PubMed]

56. Yuan, X.; Calatayud, V.; Jiang, L.; Manning, W.J.; Hayes, F.; Tian, Y.; Feng, Z. Assessing the effects of ambient ozone in China on snap bean genotypes by using ethylenediurea (EDU). *Environ. Pollut.* **2015**, *205*, 199–208. [CrossRef] [PubMed]

57. Manning, W.J.; Flagler, R.B.; Frenkel, M.A. Assessing plant response to ambient ozone: Growth of ozone-sensitive loblolly pine seedlings treated with ethylenediurea or sodium erythorbate. *Environ. Pollut.* **2003**, *126*, 73–81. [CrossRef]

58. Singh, S.; Agrawal, S.B. Use of ethylene diurea (EDU) in assessing the impact of ozone on growth and productivity of five cultivars of Indian wheat (*Triticum aestivum* L.). *Environ. Monit. Assess.* **2009**, *159*, 125. [CrossRef] [PubMed]

59. Ashrafuzzaman, M.; Lubna, F.A.; Holtkamp, F.; Manning, W.J.; Kraska, T.; Frei, M. Diagnosing ozone stress and differential tolerance in rice (*Oryza sativa* L.) with ethylenediurea (EDU). *Environ. Pollut.* **2017**, *230*, 339–350. [CrossRef] [PubMed]

60. Ueda, Y.; Uehara, N.; Sasaki, H.; Kobayashi, K.; Yamakawa, T. Impacts of acute ozone stress on superoxide dismutase (SOD) expression and reactive oxygen species (ROS) formation in rice leaves. *Plant Physiol. Biochem.* **2013**, *70*, 396–402. [CrossRef] [PubMed]

61. Hassan, I.A.; Bell, J.N.B.; Marshall, F.M. Effects of air filtration on Egyptian clover (*Trifolium alexandrinum* L. cv. Messkawy) grown in open-top chambers in a rural site in Egypt. *Res. J. Biol. Sci.* **2007**, *2*, 395–402.

62. Oksanen, E.; Rousi, M. Differences of Betula origins in ozone sensitivity based on open-field experiment over two growing seasons. *Can. J. For. Res.* **2001**, *31*, 804–811. [CrossRef]

63. Akhtar, N.; Yamaguchi, M.; Inada, H.; Hoshino, D.; Kondo, T.; Fukami, M.; Funada, R.; Izuta, T. Effects of ozone on growth, yield and leaf gas exchange rates of four Bangladeshi cultivars of rice (*Oryza sativa* L.). *Environ. Pollut.* **2010**, *158*, 2970–2976. [CrossRef]

64. Meier, S.; Tzfadia, O.; Vallabhaneni, R.; Gehring, C.; Wurtzel, E.T. A transcriptional analysis of carotenoid, chlorophyll and plastidial isoprenoid biosynthesis genes during development and osmotic stress responses in Arabidopsis thaliana. *BMC Syst. Biol.* **2011**, *5*, 77. [CrossRef]

65. Ranieri, A.; Soldatini, G. Detoxificant systems in bean plants grown in polluted air: Effects of the antioxidant EDU. *Mediterr. Agric.* **1995**, *125*, 375–386.

66. Agathokleous, E.; Mouzaki-Paxinou, A.-C.; Saitanis, C.J.; Paoletti, E.; Manning, W.J. The first toxicological study of the antiozonant and research tool ethylene diurea (EDU) using a *Lemna minor* L. bioassay: Hints to its mode of action. *Environ. Pollut.* **2016**, *213*, 996–1006. [CrossRef] [PubMed]

Article

A New Wetness Index to Evaluate the Soil Water Availability Influence on Gross Primary Production of European Forests

Chiara Proietti [1,*], Alessandro Anav [2], Marcello Vitale [3], Silvano Fares [4],
Maria Francesca Fornasier [1], Augusto Screpanti [2], Luca Salvati [4], Elena Paoletti [5], Pierre Sicard [6]
and Alessandra De Marco [2]

[1] Institute for Environmental Protection and Research, ISPRA, Via Brancati 48, I–00144 Rome, Italy;
    mariafrancesca.fornasier@isprambiente.it
[2] Italian National Agency for New Technologies, Energy and the Environment (ENEA), C.R. Casaccia,
    Via Anguillarese 301, I–00123 S. Maria di Galeria, Rome, Italy; alessandro.anav@enea.it (A.A.);
    alessandra.demarco@enea.it (A.D.M.)
[3] Department of Environmental Biology, Sapienza University of Rome, Piazzale Aldo Moro 5,
    I–00185 Rome, Italy; marcello.vitale@uniroma1.it
[4] Council for Agricultural Research and Agricultural Economy Analysis (CREA)—Research Centre for
    Forestry and Wood, Viale Santa Margherita 80, 52100 Arezzo, Italy; silvano.fares@crea.gov.it (S.F.);
    luca.salvati@crea.gov.it (L.S.)
[5] Italian National Research Council, CNR, Via Madonna del Piano 10, I–50019 Sesto Fiorentino, Florence, Italy;
    augusto.screpanti@enea.it (A.S.); elena.paoletti@cnr.it (E.P.)
[6] ARGANS, 260 route du Pin Montard, 06410 Biot, France; psicard@argans.eu
*   Correspondence: chiara.proietti@isprambiente.it

Received: 19 February 2019; Accepted: 14 March 2019; Published: 19 March 2019

Abstract: Rising temperature, drought and more-frequent extreme climatic events have been predicted for the next decades in many regions around the globe. In this framework, soil water availability plays a pivotal role in affecting vegetation productivity, especially in arid or semi-arid environments. However, direct measurements of soil moisture are scarce, and modeling estimations are still subject to biases. Further investigation on the effect of soil moisture on plant productivity is required. This study aims at analyzing spatio-temporal variations of a modified temperature vegetation wetness index (mTVWI), a proxy of soil moisture, and evaluating its effect on gross primary production (GPP) in forests. The study was carried out in Europe on 19 representative tree species during the 2000–2010 time period. Results outline a north–south gradient of mTVWI with minimum values (low soil water availability) in Southern Europe and maximum values (high soil water availability) in Northeastern Europe. A low soil water availability negatively affected GPP from 20 to 80%, as a function of site location, tree species, and weather conditions. Such a wetness index improves our understanding of water stress impacts, which is crucial for predicting the response of forest carbon cycling to drought and aridity.

Keywords: Crown defoliation; drought; Gross Primary Production; modified Temperature Vegetation Wetness Index; MODIS; Soil Moisture

---

1. Introduction

Climate change results in rising air temperature and changes in the frequency and distribution of precipitation [1]. More-frequent extreme events are projected for the next decades in many regions around the globe [2]. Earlier studies have suggested a significant increase in frequency and severity of droughts [1,3], especially in the Mediterranean region [4–9]. Knowledge of mechanisms through which climate variations affect forest ecosystem processes, such as productivity, phenology, regeneration, and mortality, is still partial [10–12]. Consequently, there is an urgent need to realize a permanent assessment of forest sensitivity to climate change [13,14].

Mediterranean forests are considered particularly susceptible to future warming and drought conditions due to their location within the transition zone between arid and temperate climates [2,15–17]. Several studies have reported increased plant mortality rates and die-off events, reduced seedling recruitment, long-term shifts in vegetation composition, reduced radial growth, and increased crown defoliation responses [18–22]. Similarly, studies carried out in boreal forests, located in regions less constrained by drought, have reported reductions in productivity and widespread increases in tree mortality because of increasing drought stress [23]. On the one hand, warming conditions might promote drought stress, reducing gross primary production (GPP, representing the sum of gross carbon uptake by plant photosynthesis) and enhancing root and tree mortality [24–27]. On the other hand, warming conditions could increase: (i) forest growth from CO_2 fertilization, (ii) water use efficiency, and (iii) growing seasons duration [28–32].

Soil moisture plays a pivotal role in vegetation dynamics [8,33], considering that soil water availability is a crucial limiting factor for plant photosynthesis [34,35]. However, soil water content (SWC) is a critical variable depending on many factors such as precipitation amount, the duration and intensity of drought, level of soil erosion, surface runoff impacts, and rate of evapotranspiration [36,37]. Consequently, SWC is difficult to estimate both with field measurements and model estimations [38]. There are different methods to estimate SWC, such as volumetric and gravimetric methodologies [39], or remote sensing techniques [40,41]. The Soil Moisture and Ocean Salinity (SMOS) mission conducted by the European Space Agency (ESA) has the capacity to estimate soil moisture globally and routinely based on a microwave radiometer system [42]. To date, SMOS soil moisture data have provided particularly suitable long-term records of global soil moisture, even if limited to shallow soil [43].

Remote sensing can be used to generate a vast amount of information about the Earth's surface and can be useful for suitable land-management planning. A strong relationship was found between SWC and values derived from surface temperature–vegetation index (Ts–VI) methods using optical and thermal remote sensing data as inputs [44]. Based on this method, a temperature–vegetation wetness index (TVWI) was developed to estimate SWC in a forest-dominated and topographically variable landscape in eastern Canada [45]. The integration of the normalized difference vegetation index (NDVI) and potential surface temperatures (θs) was used for an indirect assessment of soil moisture [45]. Following the concepts described above, a modified version of the temperature vegetation wetness index (mTVWI) was proposed using the complementary value of crown defoliation instead of NDVI [46]. In Europe, tree crown defoliation is monitored by the International Cooperative Programme on Assessment and Monitoring of Air Pollution Effects on Forests (ICP Forests) and is regarded as morphological damage affecting the green biomass of plants [47].

The driving hypothesis in this study is that a large variation of GPP, in response to soil water availability, depends on species, location, and weather characteristics, helping to better understand the relationship between GPP and mTVWI. Using the mTVWI, and considering GPP as a proxy for photosynthesis at the canopy level, we analyzed spatio-temporal variations of mTVWI across Europe, quantifying the impact of mTVWI and thus soil water availability on the GPP of European forests.

The analysis was carried out over the time period 2000–2010 for 19 representative tree species distributed across four bio-geographic areas using a MODIS GPP at 1 km of spatial resolution (Collection5 MOD17).

2. Materials and Methods

2.1. Data Collection

Crown defoliation data were provided by the ICP Forests network (Level I monitoring plots) for the main European tree species over the time period 2000–2010. In Europe, crown condition is the most widely applied indicator for forest health and vitality assessments [43–46,48–51]. Crown defoliation of the dominant tree species is visually estimated as the relative amount of foliage (expressed as percentage) in the tree crown compared to a reference tree of the same species with full foliage, growing in the same stand and site conditions [52,53]. Crown defoliation is estimated annually for 1623 monitoring plots of Level I according to the Manual on methods and criteria for harmonised sampling, assessment, monitoring, and analysis of the effects of air pollution on forests [53]. Mean annual values of crown defoliation were calculated for each Level I monitoring plot and for all considered species as reported in Figure S1 in the supplementary material. In particular, we considered nineteen tree species: six conifers (*Larix decidua* Mill., *Picea abies* L., *Pinus halepensis* Mill., *Pinus nigra* J.F. Arnold, *Pinus pinaster* Aiton, and *Pinus sylvestris* L.), twelve broad-leaf deciduous trees (*Acer pseudoplatanus* L., *Alnus glutinosa* Gaertn, *Betula pendula* Roth, *Betula pubescens* Ehrhart, *Carpinus betulus* L., *Castanea sativa* Miller, *Fagus sylvatica* L., *Fraxinus excelsior* L., *Populus tremula* L., *Quercus petraea* Liebl., *Quercus pubescens* Willd., and *Quercus robur* L.) and one broad-leaf evergreen species (*Quercus ilex* L.) distributed across four European bio-geographic areas (Boreal, Atlantic–Temperate, Continental–Temperate, Mediterranean). Furthermore, the crown defoliation database included site characteristics such as latitude, longitude, and altitude.

Mean annual temperature data (T mean, °C) were obtained from the European Climate Assessment & Dataset (ECA&D) for the time period 2000–2010. The ECA&D consists of a daily series of observations collected at several meteorological stations (>2000) throughout Europe. The nearest meteorological station to each ICP Forests site was used to provide temperature data.

GPP data (gC m^{-2} day^{-1}) were provided by MOD17 product of the MODIS (MODerate Resolution Imaging Spectroradiometer) instrument aboard the Terra satellite, as modified by [54]. MOD17 is the first continuous satellite-driven dataset monitoring global vegetation productivity at 1 km of spatial resolution, providing 8-day, monthly, and annual GPP averages. The GPP component of the algorithm is estimated by a light use efficiency model and requires daily inputs of incoming photosynthetically active radiation (PAR) and climate variables [55–57]. The Collection5 MOD17 products, temperature, and crown defoliation data were used with a temporal resolution of 1 year, over the time period 2000–2010. All analyses were separately performed for all considered years in the 1623 plots.

2.2. Modified Temperature Vegetation Wetness Index (mTVWI)

The mTVWI [45,46] was defined as an indirect assessment of soil moisture related to NDVI and potential surface temperatures (θs). NDVI is one of the most extensively applied vegetation indices, closely associated to leaf area index (LAI), primary production [58], and the biochemical processes underlying primary production (photosynthetic activity). Following Vitale et al. (2014) [46], we used the complementary values of crown defoliation Γ, defined as 100 minus defoliation values as a proxy of NDVI.

$$\Gamma = 100 - \text{Defoliation}_{\text{plot}} \tag{1}$$

where Defoliation$_{\text{plot}}$ is the mean crown defoliation value in one specific plot (in %). Γ was calculated for all considered plots and years. The potential surface temperature (θs) was calculated in two steps, as in [45]. The first step concerned the calculation of atmospheric pressure (p, in kPa), given by:

$$p = 101.3 \left[\frac{293 - 0.0065z}{293} \right]^{5.26} \tag{2}$$

where z (in m) is the plot elevation relative to sea level. This equation is based on a simplified form of the ideal gas law for a neutrally-stratified atmosphere and a temperature of 20 °C at a standard atmosphere (i.e., 101.3 kPa).

Then, surface potential temperature θs (in K) was calculated as:

$$\theta s = T_s \left[\frac{p_0}{p} \right]^{R/c_p} \tag{3}$$

where T_S is the mean annual temperature (in K), p_0 is the average pressure at mean sea level (set at 100.0 kPa), R is the gas constant (i.e., 287 J kg^{-1} K^{-1}), and c_p is the specific heat capacity of air (~1004 J kg^{-1} K^{-1}). A scatter plot between θs and Γ was generated to derive two parameters, θ_{dry} and θ_{wet} (Figure 1), for each considered year. θ_{dry} (in K) indicates low levels of water availability limiting evapotranspiration; it was calculated as the straight line passing through points A and B, where A is the point with coordinates (max Γ; max θs in max Γ) and B is the point with coordinates (max Γ in the 99th percentile of θs; 99th percentile of θs) as depicted in Figure 1 and Figure S2.

θ_{wet} (in K) indicates an unlimited amount of water available to sustain evapotranspiration, and was calculated as the 1st percentile of θs. To derive the percentiles we considered all analyzed species together (1623 monitoring plots) for each year.

Finally, we calculated the mTVWI as follows:

$$\text{mTVWI} = \frac{\theta_{dry} - \theta_s}{\theta_{dry} - \theta_{wet}} \tag{4}$$

where mTVWI is a dimensionless quantity varying from 0 (the case where water is not available for evapotranspiration, θ_{dry}) and 1 (the case where water is freely available for evapotranspiration, θ_{wet}). Consequently, the mTVWI is an index strictly linked to forest crown defoliation (θ_{dry}), surface potential temperature, and vapor pressure deficit (VPD) conditions [46].

All the above parameters were calculated on an annual basis during the time period 2000–2010.

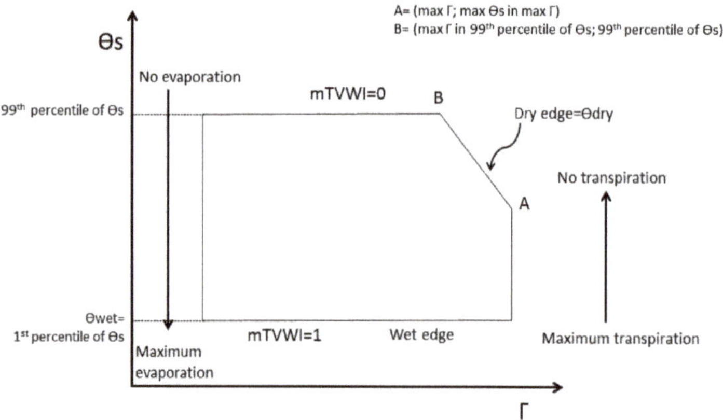

Figure 1. mTVWI scheme based on the relationship between potential surface temperature (θs) and the complementary values of crown defoliation (Γ).

2.3. Statistical Analyses

Under the assumption that partial correlation is a measure of association between two variables controlling or adjusting the effect of one or more additional variables [59], partial correlation coefficients define the degree of existing correlation between two variables when the effects of one or more variables are removed [59]. Since temperature strongly affects GPP and mTVWI, possibly causing confounding effects [60–62], we evaluated the partial correlation between annual data of GPP and the mTVWI after that temperature effect was removed [63] over the time period 2000–2010. We also performed a pair-wise Spearman' rank correlation analysis using r_s coefficients [64], a non-parametric test to investigate both linear and non-linear correlations [59,60,65,66]. Significance was tested at $p < 0.05$ after Bonferroni's correction for multiple comparisons [67]. One-way analysis of variance (ANOVA) was used to determine if there was at least one significant difference among the means of three or more independent groups. Specifically, we hypothesized that the means of mTVWI and GPP reduction in broad-leaf deciduous species significantly differed from mTVWI and GPP reduction in conifer and broad-leaf evergreen species. We used the Tukey post hoc test to find significant patterns and/or relationships ($p < 0.05$) between subgroups of broad-leaf deciduous, conifer, and broad-leaf evergreen species.

2.4. Impact of mTVWI on GPP

The impact of mTVWI on GPP was estimated similarly to the method proposed by [62,68] to quantify the impact of tropospheric ozone on GPP. The quantification of mTVWI influence on GPP (GPP related to water availability, GPP_{Water}) was expressed as follows:

$$GPP_{Water} = GPP \times mTVWI \tag{5}$$

where GPP is the original GPP derived from the MODIS product.

Finally, we calculated the percentage variation of the original GPP in respect to GPP_{Water} to evaluate the impacts of aridity on European tree species considered in this study for each year from 2000 to 2010.

2.5. Anomalies Analysis

The term anomaly is defined here as the difference between the value of a variable (e.g., GPP and mTVWI) for a single temporal period (year) and the grand mean of the same variable over a given time period (2000–2010). The analysis of anomalies was performed for each analyzed tree species, displaying GPP and mTVWI anomalies together in order to evaluate a recurrent pattern and to identify all deviations from the grand mean.

3. Results

3.1. Space–Time Distribution Patterns of mTVWI in Europe

The spatial distribution of the mTVWI values for the 19 plant species over the time period 2000–2010, is shown in Figure 2A–C. A north–south gradient in Europe is highlighted. Minimum mTVWI values (0–0.20) were mainly found in Southern Europe (Italy and Spain) while the maximum (0.80–1) were observed in Northeastern Europe (Belarus and Norway). In Central Europe, we found mTVWI values ranging from 0.40 to 0.80. Mediterranean evergreen oak (*Q. ilex* L.) showed a low mTVWI value (0.17) whilst *B. pubescens*, a typical Northern European species growing in wet conditions, showed a higher mTVWI value (0.87). In general, we found that broad-leaf deciduous species showed higher mTVWI values than conifers and the broad-leaf evergreen (Table S1). The analysis of variance (ANOVA) corroborated this evidence. In fact, the ANOVA and Tukey post hoc test indicated significant differences ($p < 0.05$) among mean mTVWI values referring to conifers, broad-leaf deciduous, and evergreen species for all years (Figure S3, Table S2). Taking into account temporal variations, Figure 3 shows the mean annual mTVWI values (±standard error, SE) over the time period 2000–2010 when considering all tree species together. We found the lowest mTVWI values in 2007 and 2008 (0.49 ± 0.01 and 0.33 ± 0.01, respectively) whilst in 2000, 2002, and 2009 we found mTVWI values ranging from 0.60 to 0.63.

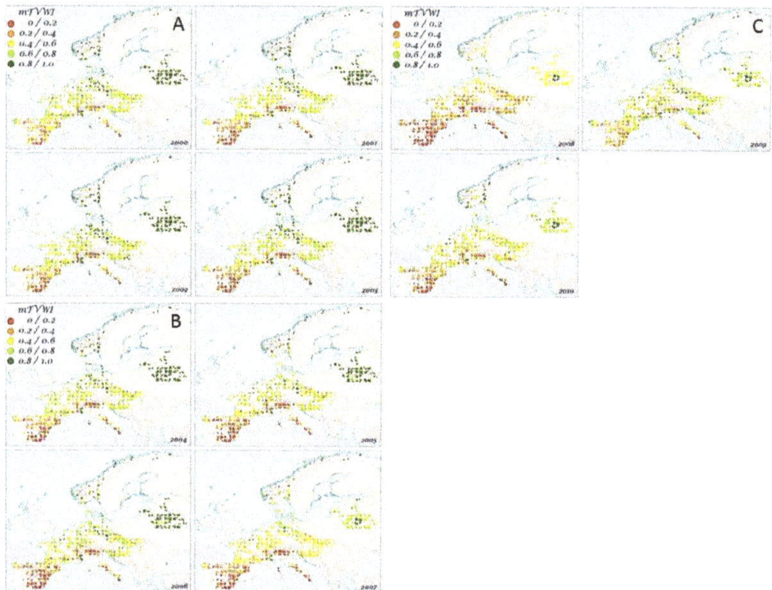

Figure 2. Spatial distribution of the mTVWI values over the time period 2000–2010. The points show ICP Forests level I sites.

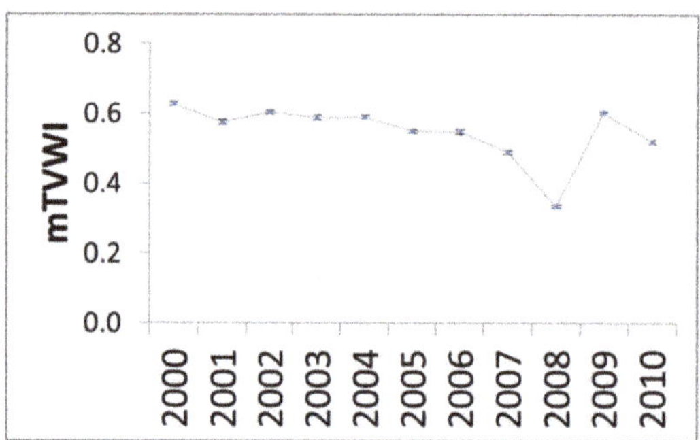

Figure 3. mean annual values of mTVWI (±SE) over the time period 2000–2010.

3.2. Correlation between GPP and mTVWI

Partial correlation coefficients between GPP and mTVWI were significant ($p < 0.05$) in most of the European tree species (Table 1). In general, GPP and mTVWI correlated positively in conifers and broad-leaf evergreen species and negatively in broad-leaf deciduous species. We found positive correlation coefficients in *L. decidua*, *P. pinaster*, *P. nigra*, *B. pendula*, and *Q. ilex* ranging from 0.22 to 0.56 ($p < 0.05$). By contrast, *P. sylvestris*, *A. pseudoplatanus*, *A. glutinosa*, *C. betulus*, *F. excelsior*, *P. tremula*, *Q. pubescens*, and *Q. robur* showed negative coefficients varying from −0.24 to −0.56 ($p < 0.05$). Other species such as *P. abies*, *P. halepensis*, *B. pubescens*, *C. sativa*, *F. sylvatica*, and *Q. petraea* did not show significant relationships between GPP and mTVWI. Mediterranean species such as *P. pinaster* and *Q. ilex* showed the highest positive correlation coefficients between GPP and mTVWI (0.36–0.50 and 0.24–0.33, respectively). Conversely, species growing in wet conditions like *P. sylvestris* and *P. tremula* showed negative correlation coefficients between GPP and mTVWI.

Table 1. Partial correlation coefficients (controlling for: temperature (T)) for the correlation between gross primary production (GPP) and the modified temperature vegetation wetness index (mTVWI). Bold indicates significance at $p < 0.05$.

SPECIES	r (GPP, mTVWI)										
	2000	2001	2002	2003	2004	2005	2006	2007	2008	2009	2010
Conifers	**0.15**	**0.15**	**0.13**	**0.09**	**0.15**	**0.13**	**0.08**	**0.12**	**0.12**	**0.10**	**0.10**
Larix decidua (N = 72)	**0.27**	**0.34**	**0.35**	0.19	**0.40**	**0.38**	**0.40**	**0.38**	0.33	**0.39**	**0.35**
Picea abies (N = 97)	−0.03	0.03	−0.04	−0.20	0.04	−0.10	−0.09	−0.02	−0.01	0.001	−0.11
Pinus halepensis (N = 82)	0.11	0.09	0.11	0.08	0.07	0.07	0.09	−0.02	0.05	0.11	0.12
Pinus nigra (N = 89)	**0.26**	**0.25**	**0.23**	0.17	0.18	**0.26**	0.15	0.16	0.17	0.21	0.15
Pinus pinaster (N = 87)	**0.46**	**0.47**	**0.46**	**0.43**	**0.45**	**0.50**	**0.44**	**0.46**	**0.48**	**0.38**	**0.36**
Pinus sylvestris (N = 77)	**−0.39**	**−0.40**	**−0.38**	**−0.38**	**−0.36**	**−0.32**	**−0.39**	**−0.39**	**−0.34**	**−0.32**	**−0.29**
Broad-leaf deciduous	**−0.20**	**−0.20**	**−0.27**	**−0.25**	**−0.22**	**−0.21**	**−0.25**	**−0.22**	**−0.18**	**−0.10**	**−0.15**
Acer pseudoplatanus (N = 100)	**−0.30**	**−0.36**	**−0.33**	**−0.37**	**−0.28**	**−0.27**	**−0.36**	**−0.32**	**−0.31**	**−0.25**	**−0.25**
Alnus glutinosa (N = 78)	**−0.44**	**−0.44**	**−0.46**	**−0.43**	**−0.43**	**−0.34**	**−0.34**	**−0.43**	−0.09	−0.09	−0.18
Betula pendula (N = 97)	**0.45**	**0.56**	**0.46**	**0.46**	**0.61**	**0.42**	**0.24**	**0.22**	0.09	0.12	−0.05
Betula pubescens (N = 71)	0.03	0.07	0.02	−0.004	−0.10	−0.21	−0.19	−0.01	−0.16	−0.003	−0.13
Carpinus betulus (N = 92)	**−0.25**	**−0.36**	**−0.30**	**−0.41**	**−0.34**	**−0.32**	**−0.26**	**−0.29**	−0.09	0.13	−0.15
Castanea sativa (N = 73)	0.01	−0.11	−0.04	−0.09	−0.08	−0.05	−0.07	−0.03	0.06	0.16	0.02
Fagus sylvatica (N = 82)	−0.14	−0.10	−0.12	−0.21	0.03	−0.01	−0.04	−0.15	−0.02	0.03	0.05
Fraxinus excelsior (N = 88)	**−0.27**	**−0.30**	**−0.35**	**−0.28**	**−0.30**	**−0.28**	**−0.30**	**−0.34**	**−0.29**	**−0.24**	**−0.30**
Populus tremula (N = 94)	**−0.55**	**−0.54**	**−0.56**	**−0.42**	**−0.53**	**−0.48**	**−0.38**	**−0.69**	**−0.51**	**−0.25**	−0.16
Quercus petraea (N = 85)	0.03	−0.04	−0.08	−0.13	−0.06	−0.10	−0.16	−0.06	−0.12	−0.03	−0.05
Quercus pubescens (N = 75)	−0.19	**−0.29**	**−0.32**	**−0.36**	**−0.29**	**−0.18**	−0.12	−0.20	−0.18	−0.03	−0.11
Quercus robur (N = 94)	**−0.27**	**−0.34**	**−0.33**	**−0.35**	**−0.34**	**−0.29**	**−0.35**	**−0.32**	−0.20	−0.09	**−0.23**
Broad-leaf evergreen	0.19	**0.27**	**0.24**	**0.25**	**0.28**	**0.27**	**0.33**	**0.26**	**0.28**	**0.20**	**0.19**
Quercus ilex (N = 90)	0.19	**0.27**	**0.24**	**0.25**	**0.28**	**0.27**	**0.33**	**0.26**	**0.28**	0.20	0.19

Table 2. Spearman rank-order correlation (rs) for the correlation between gross primary production (GPP) and the modified temperature vegetation wetness index (mTVWI). Bold indicates significance at $p < 0.05$.

SPECIES	r_s (GPP, mTVWI)										
	2000	2001	2002	2003	2004	2005	2006	2007	2008	2009	2010
Conifers	**0.26**	**0.28**	**0.23**	**0.16**	**0.24**	**0.29**	**0.21**	**0.26**	**0.23**	**0.22**	**0.16**
Larix decidua (N = 72)	0.12	0.17	0.16	−0.05	**0.26**	**0.24**	0.22	0.20	0.13	0.23	0.16
Picea abies (N = 97)	−0.19	−0.13	−0.17	−0.34	−0.06	−0.15	−0.16	−0.08	−0.08	−0.04	−0.22
Pinus halepensis (N = 82)	0.19	0.22	0.18	0.20	0.19	0.14	0.17	0.05	0.12	0.19	0.16
Pinus nigra (N = 89)	**0.25**	**0.22**	0.20	0.13	0.15	**0.34**	**0.23**	0.18	**0.27**	**0.31**	0.21
Pinus pinaster (N = 87)	**0.47**	**0.45**	**0.45**	**0.46**	**0.47**	**0.47**	**0.48**	**0.46**	**0.51**	**0.40**	**0.42**
Pinus sylvestris (N = 77)	−0.34	−0.38	−0.40	−0.40	−0.37	−0.36	−0.49	−0.28	−0.41	−0.48	−0.43
Broad-leaf deciduous	**−0.58**	**−0.61**	**−0.66**	**−0.63**	**−0.65**	**−0.57**	**−0.63**	**−0.59**	**−0.53**	**−0.43**	**−0.44**
Acer pseudoplatanus (N = 100)	−0.43	−0.54	−0.49	−0.52	−0.39	−0.44	−0.47	−0.38	−0.44	−0.39	−0.35
Alnus glutinosa (N = 78)	−0.60	−0.48	−0.67	−0.54	−0.63	−0.52	−0.59	−0.61	−0.40	−0.22	−0.21
Betula pendula (N = 97)	−0.39	−0.27	−0.54	−0.42	−0.45	−0.42	−0.48	−0.35	−0.26	−0.26	−0.08
Betula pubescens(N = 71)	−0.56	−0.55	−0.50	−0.50	−0.52	−0.48	−0.53	−0.60	−0.60	−0.44	−0.36
Carpinus betulus (N = 92)	−0.35	−0.49	−0.48	−0.40	−0.43	−0.26	−0.39	−0.30	−0.13	0.05	−0.20
Castanea sativa (N = 73)	−0.15	−0.32	−0.28	−0.29	−0.34	−0.20	−0.20	−0.19	−0.15	0.04	−0.22
Fagus sylvatica (N = 82)	−0.56	−0.59	−0.59	−0.73	−0.49	−0.47	−0.42	−0.56	−0.50	−0.30	−0.42
Fraxinus excelsior (N = 88)	−0.49	−0.52	−0.56	−0.53	−0.59	−0.47	−0.52	−0.55	−0.48	−0.46	−0.44
Populus tremula (N = 94)	−0.45	−0.34	−0.44	−0.29	−0.32	−0.43	−0.45	−0.31	−0.19	−0.18	−0.11
Quercus petraea (N = 85)	0.01	−0.22	−0.19	−0.21	−0.24	−0.19	−0.26	−0.09	−0.17	−0.08	−0.07
Quercus pubescens (N = 75)	−0.25	−0.31	−0.37	−0.43	−0.39	−0.25	−0.20	−0.20	−0.18	−0.07	−0.09
Quercus robur(N = 94)	−0.35	−0.51	−0.54	−0.53	−0.57	−0.47	−0.57	−0.38	−0.44	−0.33	−0.35
Broad-leaf evergreen	**0.23**	**0.22**	**0.24**	**0.25**	**0.22**	**0.21**	**0.24**	**0.21**	0.20	0.09	0.13
Quercus ilex (N = 90)	**0.23**	**0.22**	**0.24**	**0.25**	**0.22**	**0.21**	**0.24**	**0.21**	0.20	0.09	0.13

Spearman rank-order correlation coefficients between GPP and mTVWI (Table 2) were in agreement with partial correlation analysis results. The GPP and mTVWI relationships were positive and statistically significant ($p < 0.05$) in conifers and broad-leaf evergreen species and negative in broad-leaf deciduous species. We found positive correlation coefficients in *L. decidua*, *P. nigra*, *P. pinaster*, and *Q. ilex* (the same species highlighted in the partial correlation analysis except for *B. pendula*) and negative coefficients in most of the analyzed species (e.g., *P. sylvestris*, *A. pseudoplatanus*, *B. pubescens*, *F. excelsior*).

3.3. Impact of mTVWI on GPP

The percentage change of GPP due to mTVWI is shown in Figure 4A–C. The largest decrease occurred in Mediterranean countries, whereas the smallest decrease occurred in Belarus and Norway for all considered years. Italy and Spain showed a GPP reduction ranging from 40 to 100%. Central Europe presented a GPP reduction ranging from 20 to 40% for all years except for 2007 and 2008, where the GPP reduction reached 80%. Northern and Northeastern Europe were characterized by a small GPP reduction (0–40%). Table 3 shows a plant species-based GPP reduction (%) for all years. Conifers showed a mean GPP reduction of nearly 50% ranging from 40% in *P. sylvestris* to 80% in *P. halepensis*, *P. nigra*, and *P. pinaster*. The broad-leaf deciduous species showed an average GPP reduction close to 40%, with some species showing higher reductions (54%, *F. sylvatica* and *Q. pubescens*) and other species with limited GPP reduction (20% *B. pubescens*, *B. pendula*, *P. tremula*). By contrast, *Q. ilex* presented values of GPP reduction close to 70% for all considered years. Figures 5–9 show the average GPP reduction values over the time period 2000–2010 (standard deviations are shown in Figure S4). The largest decrease occurred in *F. sylvatica*, *P. halepensis*, *P. nigra*, *P. pinaster*, and *Q. ilex*, whereas the smallest decrease occurred in *B. pendula* and *B. pubescens*.

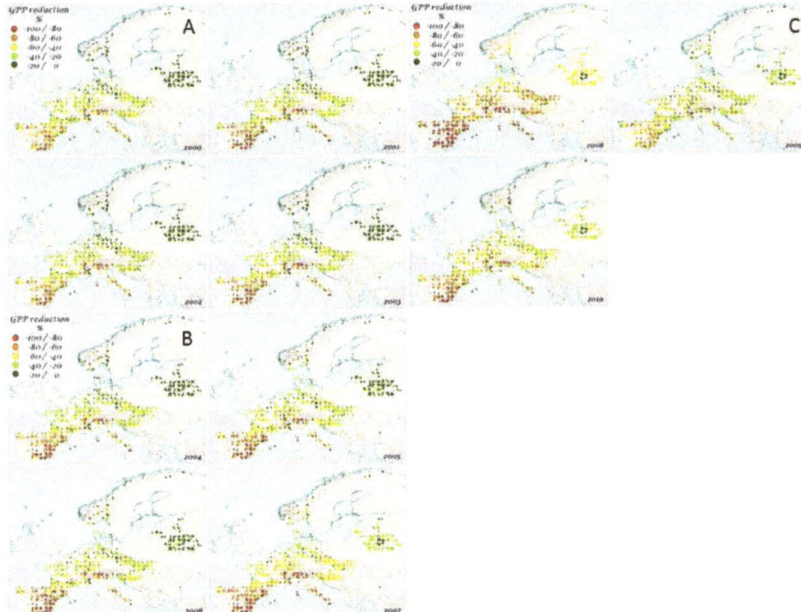

Figure 4. The percentage change of gross primary production (GPP) due to mTVWI, a proxy of soil water availability, over the time period 2000–2010.

Table 3. Species-specific GPP reduction (%) due to mTVWI over the time period 2000–2010 (±standard deviation).

SPECIES	GPP Reduction (%)										
	2000	2001	2002	2003	2004	2005	2006	2007	2008	2009	2010
Conifers	−49.3 ± 21.88	−56.9 ± 25.21	−52.70 ± 23.57	−55.53 ± 24.95	−55.61 ± 25.27	−59.55 ± 25.31	−60.16 ± 26.51	−63.76 ± 20.52	−75.96 ± 15.19	−48.34 ± 19.79	−58.28 ± 20.57
Larix decidua (N = 72)	−47.03 ± 26.80	−52.55 ± 30.60	−49.39 ± 29.58	−52.17 ± 30.87	−51.48 ± 30.86	−54.92 ± 30.29	−53.94 ± 29.90	−61.32 ± 25.85	−73.36 ± 18.01	−44.48 ± 22.74	−54.16 ± 22.50
Picea abies (N = 97)	−46.81 ± 21.14	−53.01 ± 23.95	−50.89 ± 22.88	−52.73 ± 24.68	−52.40 ± 24.11	−57.16 ± 24.67	−57.40 ± 26.80	−62.71 ± 19.87	−74.80 ± 15.04	−48.39 ± 23.29	−55.68 ± 20.79
Pinus halepensis (N = 82)	−60.97 ± 13.87	−72.27 ± 15.21	−64.11 ± 15.04	−69.11 ± 13.14	−70.40 ± 14.33	−71.42 ± 15.97	−72.19 ± 17.94	−74.34 ± 11.54	−83.21 ± 9.55	−51.95 ± 13.39	−67.27 ± 16.19
Pinus nigra (N = 89)	−57.18 ± 19.34	−66.86 ± 21.72	−60.91 ± 20.76	−65.13 ± 21.87	−65.46 ± 22.80	−69.38 ± 23.43	−69.50 ± 25.00	−71.85 ± 18.04	−80.98 ± 14.90	−54.54 ± 19.29	−64.51 ± 20.11
Pinus pinaster (N = 87)	−49.86 ± 21.08	−58.87 ± 22.29	−54.76 ± 21.36	−57.23 ± 21.51	−56.95 ± 22.31	−62.92 ± 22.41	−65.48 ± 23.40	−62.71 ± 18.68	−77.25 ± 13.88	−51.19 ± 17.28	−62.38 ± 18.85
Pinus sylvestris (N = 77)	−32.52 ± 16.28	−36.38 ± 20.16	−34.10 ± 18.45	−34.71 ± 20.37	−34.96 ± 19.87	−39.02 ± 19.79	−39.88 ± 21.36	−47.88 ± 16.70	−64.88 ± 11.76	−37.46 ± 15.50	−44.0915.62
Broad-leaf deciduous	−30.13 ± 17.64	−33.47 ± 20.71	−31.57 ± 19.15	−32.25 ± 21.08	−31.87 ± 20.26	−35.97 ± 20.44	−36.09 ± 21.42	−43.21 ± 19.37	−60.62 ± 17.51	−34.71 ± 17.31	−41.54 ± 18.45
Acer pseudoplatanus (N = 100)	−39.80 ± 16.70	−43.04 ± 19.07	−41.37 ± 18.60	−42.05 ± 19.07	−40.33 ± 19.43	−43.62 ± 22.10	−45.78 ± 23.62	−50.00 ± 18.88	−67.49 ± 15.08	−41.71 ± 20.24	−46.72 ± 18.06
Alnus glutinosa (N = 78)	−23.37 ± 16.78	−25.45 ± 19.36	−23.83 ± 17.43	−22.99 ± 19.36	−23.79 ± 18.10	−28.17 ± 17.80	−27.56 ± 18.45	−37.90 ± 16.66	−58.46 ± 12.43	−30.27 ± 11.82	−38.93 ± 12.90
Betula pendula (N = 97)	−19.16 ± 14.85	−20.29 ± 16.48	−19.71 ± 14.01	−17.57 ± 17.24	−18.65 ± 16.06	−23.91 ± 15.97	−22.69 ± 15.05	−25.51 ± 21.38	−40.87 ± 28.07	−20.91 ± 15.69	−26.73 ± 19.35
Betula pubescens (N = 71)	−14.70 ± 8.83	−13.15 ± 8.81	−13.21 ± 8.49	−13.61 ± 8.86	−15.21 ± 9.29	−19.95 ± 10.49	−19.36 ± 10.92	−26.56 ± 9.98	−51.50 ± 8.50	−23.63 ± 9.12	−27.30 ± 23.74
Carpinus betulus (N = 92)	−34.32 ± 10.33	−35.58 ± 10.93	−32.82 ± 11.04	−34.92 ± 11.65	−33.04 ± 10.87	−37.06 ± 10.69	−40.06 ± 14.15	−46.35 ± 9.34	−64.33 ± 9.12	−41.71 ± 15.11	−48.02 ± 12.63
Castanea sativa (N = 73)	−37.85 ± 19.07	−44.29 ± 21.25	−41.53 ± 20.94	−43.46 ± 20.73	−42.68 ± 21.29	−47.42 ± 23.40	−48.30 ± 25.22	−52.09 ± 18.51	−67.89 ± 15.11	−44.45 ± 21.08	−50.65 ± 20.11
Fagus sylvatica (N = 82)	−50.00 ± 17.73	−55.39 ± 22.99	−53.18 ± 20.40	−55.21 ± 22.71	−53.60 ± 23.53	−58.35 ± 22.39	−58.94 ± 22.73	−64.32 ± 18.64	−75.85 ± 15.85	−49.01 ± 19.95	−58.94 ± 20.43
Fraxinus excelsior (N = 88)	−28.83 ± 15.02	−31.53 ± 16.76	−30.12 ± 16.06	−30.41 ± 16.85	−30.13 ± 16.09	−35.10 ± 17.82	−34.70 ± 17.48	−42.52 ± 13.44	−61.23 ± 11.33	−36.98 ± 15.12	−41.48 ± 13.84
Populus tremula (N = 94)	−16.82 ± 13.54	−18.97 ± 16.69	−18.38 ± 15.56	−16.13 ± 17.36	−17.38 ± 16.26	−21.96 ± 15.41	−20.94 ± 16.51	−30.15 ± 16.07	−52.00 ± 18.05	−26.42 ± 13.37	−34.31 ± 14.00
Quercus petraea (N = 85)	−30.58 ± 10.98	−34.87 ± 12.75	−32.67 ± 11.30	−34.48 ± 12.66	−33.18 ± 12.63	−36.39 ± 12.74	−36.91 ± 13.19	−45.67 ± 10.94	−62.12 ± 8.50	−33.61 ± 9.81	−41.08 ± 9.14
Quercus pubescens (N = 75)	−42.77 ± 13.46	−53.33 ± 16.52	−47.00 ± 15.40	−52.05 ± 16.14	−50.81 ± 16.49	−52.35 ± 16.88	−50.06 ± 18.09	−60.44 ± 14.63	−70.28 ± 10.76	−37.86 ± 11.42	−48.77 ± 13.50
Quercus robur (N = 94)	−25.50 ± 10.64	−29.02 ± 11.41	−27.54 ± 10.87	−28.07 ± 12.46	−27.62 ± 11.86	−31.21 ± 12.19	−31.02 ± 13.02	−41.24 ± 10.72	−59.25 ± 10.41	−31.63 ± 13.41	−37.82 ± 10.97
Broad-leaf evergreen	−57.63 ± 14.65	−71.26 ± 15.08	−64.20 ± 15.15	−68.88 ± 14.72	−69.85 ± 15.62	−73.12 ± 16.53	−72.16 ± 18.08	−74.46 ± 12.82	−83.41 ± 10.01	−54.08 ± 12.52	−67.98 ± 15.86
Quercus ilex (N = 90)	−57.63 ± 14.65	−71.26 ± 15.08	−64.20 ± 15.15	−68.88 ± 14.72	−69.85 ± 15.62	−73.12 ± 16.53	−72.16 ± 18.08	−74.46 ± 12.82	−83.41 ± 10.01	−54.08 ± 12.52	−67.98 ± 15.86

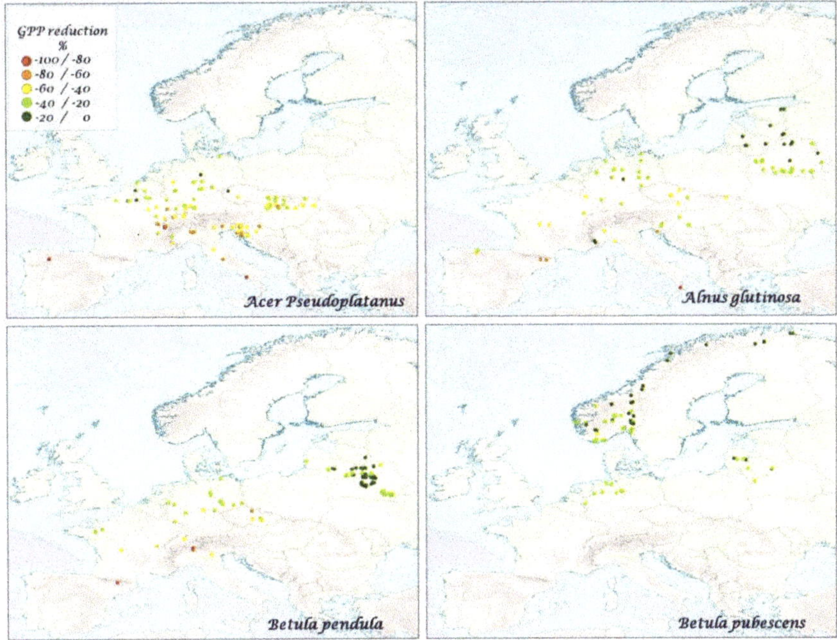

Figure 5. The mean species–specific GPP reduction (%) for *A. pseudoplatanus*, *A. glutinosa*, *B. pendula*, *B. pubescens* over the time period 2000–2010.

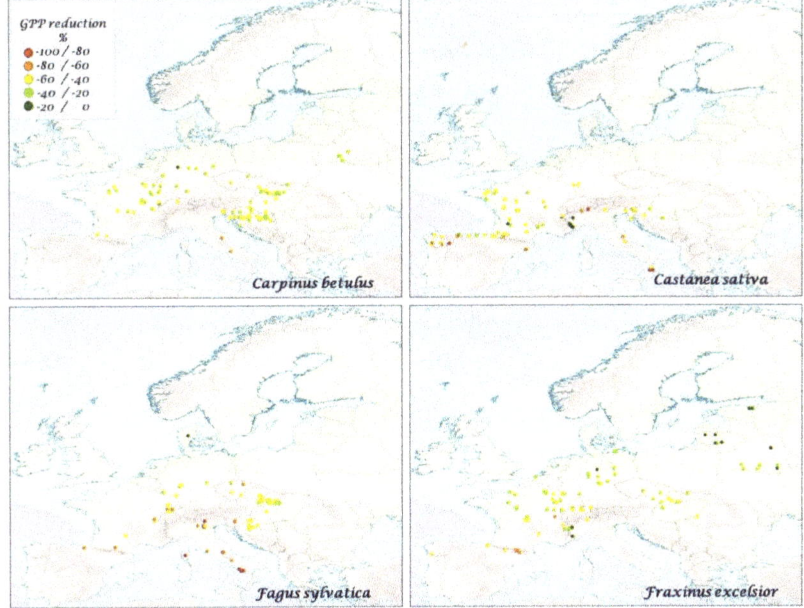

Figure 6. The mean species–specific GPP reduction (%) for *C. betulus*, *C. sativa*, *F. sylvatica*, *F. excelsior* over the time period 2000–2010.

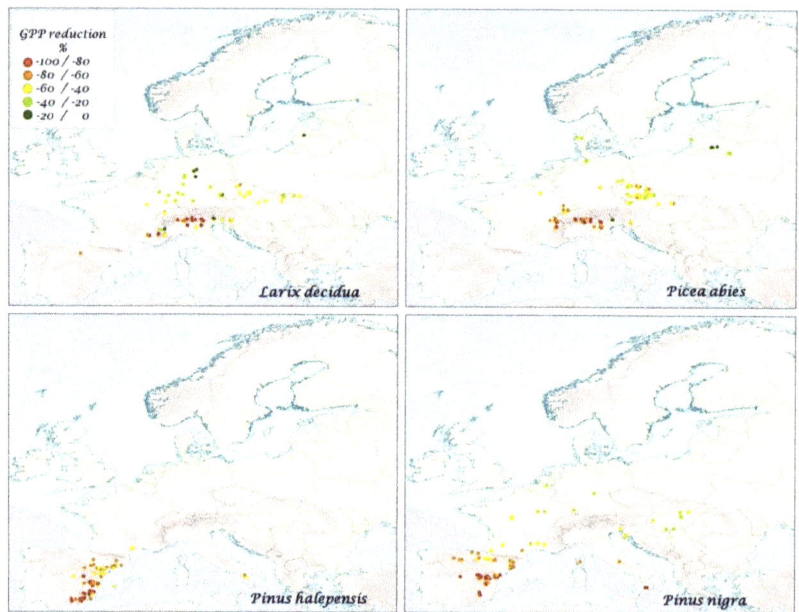

Figure 7. The mean species–specific GPP reduction (%) for *L. decidua, P. abies, P. halepensis, P. nigra* over the time period 2000–2010.

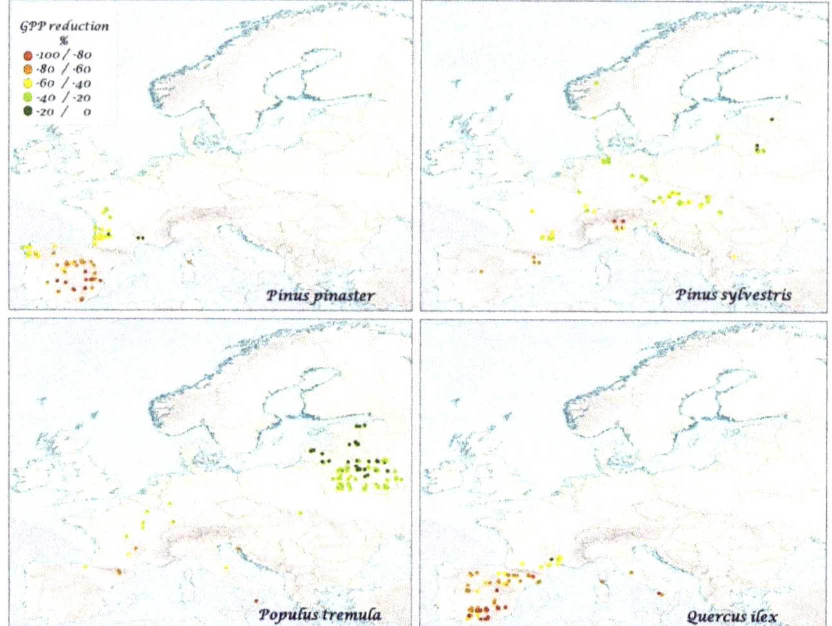

Figure 8. The mean species–specific GPP reduction (%) for *P. pinaster, P. sylvestris, P. tremula, Q. ilex* over the time period 2000–2010.

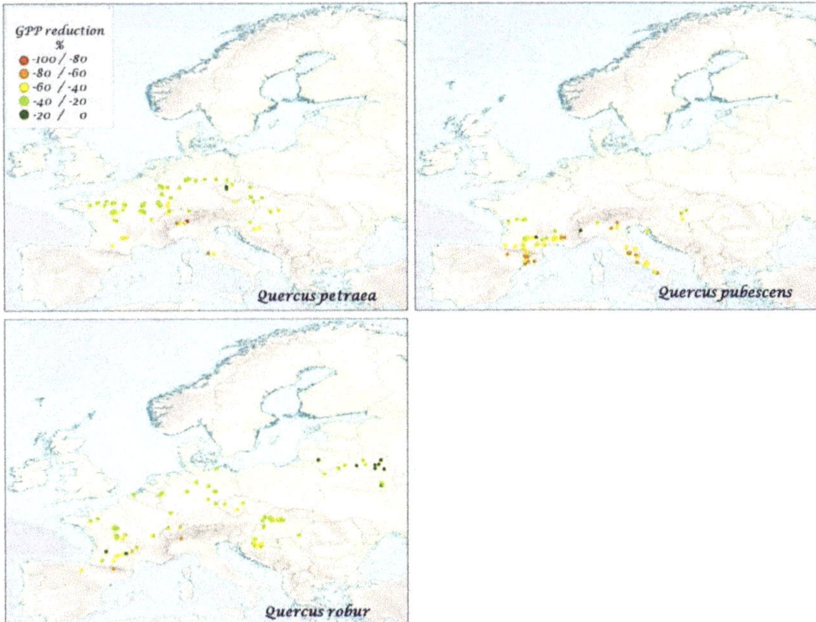

Figure 9. The mean species–specific GPP reduction (%) for *Q. petraea*, *Q. pubescens*, *Q. robur* over the time period 2000–2010.

In general, we found that broad-leaf evergreen (*Q. ilex*) and conifer species showed higher GPP reduction values than broad-leaf deciduous (Table 3), which was also confirmed by analysis of variance (ANOVA) and Tukey post hoc tests (Figure S5, Table S3). The anomalies analysis is shown in Figure 10. We found a species-specific pattern of GPP and mTVWI anomalies in the time period 2000–2010. In general, highly negative anomalies of mTVWI (a condition of severe aridity) often correspond to a reduction of GPP. Indeed, in 2008 a highly negative deviation from the grand mean corresponds to a GPP reduction in most of the considered species (e.g., *A. pseudoplatanus*, *F. sylvatica*, *P. abies*, *L. decidua*, *P. tremula*). Taking into account the 2003 heat wave that affected much of Europe from June to September [69], we found a highly negative deviation from the grand mean of GPP not matching with high negative mTVWI anomalies.

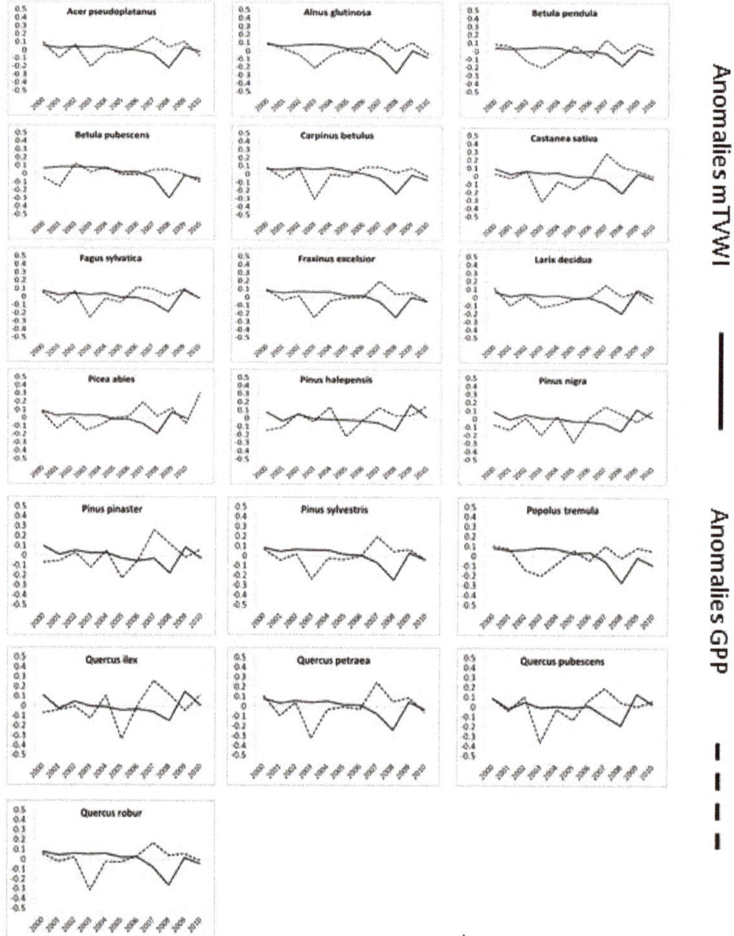

Figure 10. Anomalies of mTVWI and GPP over the time period 2000–2010. The term anomaly is defined here as the difference between the mean of mTVWI and GPP calculated for a single year and the grand–mean of mTVWI and GPP but considering the overall time period (2000–2010).

4. Discussion

4.1. Spatial and Temporal Distribution of mTVWI

Our study assessed the impact of water availability in 19 tree species distributed across four European bio-geographic areas by using a modified temperature vegetation wetness index (mTVWI). The study focused on the impact of mTVWI, a proxy for soil water availability, on GPP. Our analyses revealed a north–south gradient of the mTVWI with minimum values in Southern Europe and maximum values in Northeastern Europe over the time period 2000–2010. Precipitation over the Northern Europe increased between 10 and 40% in the 20th century [70], whereas some parts of Southern Europe dried by up to 20% [70], reporting a geographic pattern consistent with our results. However, a north–south gradient of the mTVWI is not surprising as soil moisture and soil water availability affect vegetation growth and productivity especially in arid or semi-arid ecosystems [34]. In fact, the impact of water deficit on growth is much higher in arid and semi-arid sites in which water availability principally affects the main physiological processes of vegetation (growth, photosynthesis, carbon, and nitrogen use) than in sites with a moderate or well-balanced water supply [71–74]. A significant decrease of the mTVWI (−40% respect to others years) was observed in 2007 and 2008 over Europe. This reduction is likely related to high values of crown defoliation [75], especially in Mediterranean areas, impacting mTVWI values. Unfortunately, due to the lack of measurements of soil water content into the field conditions in our sites, it is unfeasible to compare the estimation of the index with real world measurements.

4.2. Correlations, Impacts, and Anomalies of GPP and mTVWI

Significant correlations between GPP and mTVWI were found in most of the European tree species. The highest positive coefficients were found for two Mediterranean species, *Q. ilex* and *P. pinaster*. Greater water availability allows higher stomatal conductance and thus a positive relationship between mTVWI and GPP [76]. Conversely, we found negative GPP–mTVWI relationships in other species (e.g., *P. sylvestris*, *P. tremula*) mainly growing in Central–Northern Europe. These results were well corroborated by Pasho et al. [77] in 2011, where they found a strong association between tree growth and a drought index (the standardized precipitation index) in *P. sylvestris*. On the other hand, other studies in Northern European countries did not find any association between SPI and crown defoliation (e.g., [78]). The different responses to drought from tree species are related to different strategies in coping with drought and local conditions of the forest stands [77]. As a function of site location and tree species, mTVWI negatively affected GPP from 20 to 80%. We found that broad-leaf evergreen (*Q. ilex*) and conifer species showed higher GPP reduction than broad-leaf deciduous species. A generalized increase in defoliation and mortality occurring was observed during 1987–2007 in Southern European forests (principally in *Q. ilex* stands), largely related to a severe drought regime [79], thus confirming our results. In agreement with our results, a high sensitivity to drought was found for broad-leaf Mediterranean forests [24]. *P. pinaster*, a Mediterranean species, showed high GPP reductions (50–80%) over the years, which are in line with earlier studies suggesting a high dependence of water availability in mixed pine–oak Mediterranean forests [22,80–82]. Probably, forests located in sites experiencing strong drought events (e.g., Southern European forests) could reduce their carbon sink efficiency and contribute to the reduction in carbon sink in the Northern Hemisphere [82,83]. However, the response of GPP to water availability significantly varied among species and locations. For example, *B. pendula* and *B. pubescens*, mainly distributed in northern temperate regions [84], showed low GPP reductions (10–30%). Conversely, *F. excelsior* was characterized by moderate-to-high GPP reductions (30–60%) being a water-demanding species suffering severe damages under increasing temperature [85]. Similarly, *F. sylvatica* showed high GPP reductions (up to 70%) likely due to its high sensitivity to drought [24]. Furthermore, an increase of crown defoliation was reported for the beech, especially in the Mediterranean area [50]. In fact, sensitivity to extreme temperature in *F. sylvatica* increased with the decreasing of soil water availability in the Mediterranean area [86].

Anomalies analysis showed a similar pattern of mTVWI and GPP anomalies over the time period 2000–2010, highlighting that negative anomalies of mTVWI (more aridity) induced a reduction of GPP. This result is in line with several studies (e.g., [87,88]) that reported a reduction of GPP NPP related to drought. Taking into account the 2003 heat wave, we found a highly negative deviation from the grand mean of GPP, whereas mTVWI did not show a corresponding variation. However, it should be noted that mTVWI was calculated considering temperature of the overall year (2003), whilst the peak temperatures in 2003 were found only during the growing season [68], thus impacting mainly GPP.

5. Conclusions

Models such as CMIP5 running under a higher temperature increased the potential evapotranspiration and changes in seasonal precipitation patterns and highlighted an intensification of drought events both in frequency and intensity [1,3], especially in the Mediterranean region [5,6]. Increasing climate and soil aridity are expected to cause growth decline and enhancement of mortality particularly in drought-sensitive forest species [15,16,89–91]. Thus, understanding and predicting the impacts of climate change on ecosystems is one of the pivotal challenges for the global change science [92,93]. As climate change is strictly dependent on the increase of the emissions of air pollutants, and consequently on air pollutant concentrations [1,3], we cannot study the impacts of climate change on soil aridity without considering air pollution impacts. Our study contributes to an assessment of GPP responses to soil water availability based on a dimensionless index—the modified temperature vegetation wetness index (mTVWI)—which is an indirect measure of soil moisture tightly linked to VPD conditions. Empirical results of this study outline a negative impact of low soil water availability on GPP, although it varied according to plant functional types, years, and bio-geographic areas. In fact, tree species can vary their ability to adjust physiological functions to soil water deficit [94,95]. Moreover, mTVWI could be used to estimate soil water availability when direct measurements and/or model estimations are lacking. Thus, its application can improve quantification of the carbon gain of forests undergoing drought promptly, to inform forest management planners of short-term climate change and its influence on the status of forest health.

Supplementary Materials: The following are available online at http://www.mdpi.com/2225-1154/7/3/42/s1, Figure S1: title, Table S1: title, Video S1: title.

Author Contributions: Conceptualization, C.P., M.V., and A.D.M.; Methodology, C.P., M.V., and F.F.; Software, C.P.; Validation, C.P. and A.A.; Formal Analysis, C.P.; Investigation, C.P. and A.A.; Resources, A.D.M.; Data Curation, C.P. and A.D.M.; Writing—Original Draft Preparation, C.P.; Writing—Review & Editing, C.P., A.A., S.F., L.S.E.P., M.V., A.D.M., F.F., A.S. and P.S.; Visualization, A.S.; Supervision, A.D.M. and E.P.; Project Administration, E.P.; Funding Acquisition, E.P.

Funding: This research received no external funding

Acknowledgments: This work was carried out within the IUFRO Task Force on Climate Change and Forest Health.

Conflicts of Interest: No conflict of interest.

References

1. Dai, A. Increasing drought under global warming in observations and models. *Nat. Clim. Chang.* **2013**, *3*, 52–58. [CrossRef]
2. Huang, J.; Yu, H.; Guan, X.; Wang, G.; Guo, R. Accelerated dryland expansion under climate change. *Nat. Clim. Chang.* **2016**, *6*, 166–171. [CrossRef]
3. Vicente-Serrano, S.M.; Camarero, J.J.; Azorın–Molina, C. Diverse responses of forest growth to drought timescales in the Northern Hemisphere. *Glob. Ecol. Biogeogr.* **2014**, *23*, 1019–1030. [CrossRef]
4. Mavrakis, A.; Salvati, L. Analyzing the behavior of selected risk indexes during the 2007 Greek forest fires. *Int. J. Environ. Res.* **2015**, *9*, 831–840.
5. Gao, X.; Giorgi, F. Increased aridity in the Mediterranean region under greenhouse gas forcing estimated from high resolution simulations with a regional climate model. *Glob. Planet Chang.* **2008**, *62*, 195–209. [CrossRef]
6. Hertig, E.; Jacoboeit, J. Downscaling future climate change: Temperature scenarios for the Mediterranean area. *Glob. Planet Chang.* **2008**, *63*, 127–131. [CrossRef]
7. Salvati, L.; Zitti, M.; Ceccarelli, T.; Perini, L. Building–up a synthetic index of land vulnerability to drought and desertification. *Geogr. Res.* **2009**, *47*, 280–291. [CrossRef]
8. Salvati, L.; Perini, L.; Sabbi, A.; Bajocco, S. Climate aridity and land use changes: A regional-scale analysis. *Geogr. Res.* **2012**, *50*, 193–203. [CrossRef]
9. Colantoni, A.; Ferrara, C.; Perini, L.; Salvati, L. Assessing Trends in Climate Aridity and Vulnerability to Soil Degradation in Italy. *Ecol. Indic.* **2015**, *48*, 599–604. [CrossRef]
10. Michaletz, S.T.; Cheng, D.; Kerkhoff, A.J.; Enquist, B.J. Convergence of terrestrial plant production across global climate gradients. *Nature* **2014**. [CrossRef]
11. Cutini, A.; Manetti, M.C.; Mazza, G.; Moretti, V.; Salvati, L. Climate variability and growth rate of *Pinus pinea* L. in Castelporziano forest: An exploratory data analysis. *Rend. Accad. Nazion. Lincei* **2015**, *26*, 413–420. [CrossRef]

12. Chu, C.; Bartlett, M.; Wang, Y.; He, F.; Weiner, J.; Chave, J.; Sack, L. Does climate directly influence NPP globally? *Glob. Chang. Biol.* **2016**, *22*, 12–24. [CrossRef]

13. Salvati, R.; Salvati, L.; Ferrara, A.; Corona, P.; Barbati, A. Estimating the sensitivity to desertification of Italian forests. *iForest* **2014**, *8*, 287–294. [CrossRef]

14. Zang, C.; Meier, C.H.; Dittmar, C.; Rothe, A.; Menzel, A. Patterns of drought tolerance in major European temperate forest trees: Climatic drivers and levels of variability. *Glob. Chang. Biol.* **2014**, *20*, 3767–3779. [CrossRef]

15. Morales, P.; Hickler, T.; Rowell, D.P.; Smith, B.; Sykes, M.T. Changes in European ecosystem productivity and carbon balance driven by regional climate model output. *Glob. Chang. Biol.* **2007**, *13*, 108–122. [CrossRef]

16. Anav, A.; Mariotti, A. Sensitivity of natural vegetation to climate change in the Euro-Mediterranean area. *Clim. Res.* **2011**, *46*, 277–292. [CrossRef]

17. Barbeta, A.; Mejía-Chang, M.; Ogaya, R.; Voltas, J.; Dawson, T.E.; Penuelas, J. The combined effects of a long–term experimental drought and an extreme drought on the use of plant-water sources in a Mediterranean forest. *Glob. Chang. Biol.* **2014**, *21*, 1213–1225. [CrossRef]

18. Sicard, P.; Dalstein-Richier, L. Health and vitality assessment of two common pine species in the context of climate change in southern Europe. *Environ. Res.* **2015**, *137*, 235–245. [CrossRef]

19. Allen, C.D.; Macalady, A.K.; Chenchouni, H.; Bachelet, D.; McDowell, N.; Vennetier, M.; Kitzberger, T.; Rigling, A.; Breshears, D.D.; Hogg, E.H.; et al. A global overview of drought and heat-induced tree mortality reveals emerging climate change risks for forests. *For. Ecol. Manag.* **2010**, *259*, 660–684. [CrossRef]

20. Mueller, R.C.; Scudder, C.M.; Porter, M.E.; Talbot, T.R.; Gehring, C.A.; Whitham, T.G. Differential tree mortality in response to severe drought: Evidence for long-term vegetation shifts. *J. Ecol.* **2005**, *93*, 1085–1093. [CrossRef]

21. Breshears, D.D.; Cobb, N.S.; Rich, P.M.; Price, K.P.; Allen, C.D.; Balice, R.G.; Romme, W.H.; Kastens, J.H.; Floyd, M.L.; Belnap, J.; et al. Regional vegetation die–off in response to global-change-type drought. *Proc. Natl. Acad. Sci. USA* **2005**, *102*, 15144–15148. [CrossRef]

22. Andreu, L.; Gutiérrez, E.; Macias, M.; Ribas, M.; Bosch, O.; Camarero, J.J. Climate increases regional tree-growth variability in Iberian pine forests. *Glob. Chang. Biol.* **2007**, *13*, 804–815. [CrossRef]

23. Peng, C.; Ma, Z.; Lei, X.; Zhu, Q.; Chen, H.; Wang, W.; Liu, S.; Li, W.; Fang XZhou, X. A drought–induced pervasive increase in tree mortality across Canada's boreal forests. *Nat. Clim. Chang.* **2011**, *1*, 467–471. [CrossRef]

24. Granier, A.; Reichestein, M.; Brèda, N.; Janssens, I.A.; Falge, E.; Ciais, P.; Grünwald, T.; Aubinet, T.; Berbigier, P.; Bernhofer, C.; et al. Evidence for soil water control on carbon and water dynamics in European forests during the extremely dry year: 2003. *Agric. For. Meteorol.* **2007**, *143*, 123–145. [CrossRef]

25. Friend, A.D.; Lucht, W.; Rademacher, T.T.; Keribin, R.; Betts, R.; Cadule, P.; Ciais, P.; Clark, D.B.; Dankers, R.; Falloon, P.D.; et al. Carbon residence time dominates uncertainty in terrestrial vegetation responses to future climate and atmospheric CO_2. *Proc. Natl. Acad. Sci. USA* **2014**, *111*, 3280–3285. [CrossRef] [PubMed]

26. Allen, C.D.; Breshears, D.D.; McDowell, N.G. On underestimation of global vulnerability to tree mortality and forest die–off from hotter drought in the Anthropocene. *Ecosphere* **2015**, *6*, 1–55. [CrossRef]

27. McDowell, N.G.; Williams, A.P.; Xu, C.; Pockman, W.T.; Dickman, L.T.; Sevanto, S.; Pangle, R.; Limousin, J.; Plaut, J.; Mackay, D.S.; et al. Multi–scale predictions of massive conifer mortality due to chronic temperature rise. *Nat. Clim. Chang.* **2016**, *6*, 295–300. [CrossRef]

28. Ayres, M.P.; Lombardero, M.J. Assessing the consequences of global change for forest disturbances for herbivores and pathogens. *Sci. Total Environ.* **2000**, *262*, 263–286. [CrossRef]

29. Bachelet, D.; Neilson, R.P.; Hickler, T.; Drapek, R.J.; Lenihan, J.M.; Sykes, M.T.; Smith, B.; Sitch, S.; Thonicke, K. Simulating past and future dynamics of natural ecosystems in the United States. *Glob. Biogeochem. Cycles* **2003**, *17*, 1045. [CrossRef]

30. Lucht, W.; Schaphoff, S.; Erbrecht, T.; Heyder, U.; Cramer, W. Terrestrial vegetation redistribution and carbon balance under climate change. *Carbon Balance Manag.* **2006**, *1*, 6. [CrossRef]

31. Scholze, M.; Knorr, W.; Arnell, N.W.; Prentice, I. A climate–change risk analysis for world ecosystems. *Proc. Natl. Acad. Sci. USA* **2006**, *103*, 13116–13120. [CrossRef] [PubMed]

32. Lloyd, A.H.; Bunn, A.G. Responses of the circumpolar boreal forest to 20th century climate variability. *Environ. Res. Lett.* **2007**, *2*, 045013. [CrossRef]

33. Cleverly, J.; Eamus, D.; Restrepo Coupe, N.; Chen, C.; Maes, W.; Li, L.; Faux, R.; Santini, N.S.; Rumman, R.; Yu, Q.; et al. Soil moisture controls on phenology and productivity in a semi–arid critical zone. *Sci. Total Environ.* **2016**, *568*, 1227–1237. [CrossRef] [PubMed]

34. Nemani, R.R.; Keeling, C.D.; Hashimoto, H.; Jolly, W.M.; Piper, S.C.; Tucker, C.J.; Myneni, R.B.; Running, S.W. Climate–Driven Increases in Global Terrestrial Net Primary Production from 1982 to 1999. *Science* **2003**, *300*, 1560–1563. [CrossRef] [PubMed]

35. Murray-Tortarolo, G.; Friedlingstein, P.; Sitch, S.; Seneviratne, S.I.; Fletcher, I.; Mueller, B.; Greve, P.; Anav, A.; Liu, Y.; Ahlström, A.; et al. The dry season intensity as a key driver of NPP trends. *Geophys. Res. Lett.* **2016**, *43*, 2632–2639. [CrossRef]

36. Salvati, L.; Petitta, M.; Ceccarelli, T.; Perini, L.; Di Battista, F.; Venezian Scarascia, M.E. Italy's renewable water resources as estimated on the basis of the monthly water balance. *Irr. Drain* **2008**, *57*, 507–515. [CrossRef]

37. Moretti, V.; Di Bartolomei, R.; Sorgi, T.; Aromolo, R.; Salvati, L. Soil water deficit and climate conditions during the dry season along the coastal–inland gradient in Castelporziano Forest, central Italy. *Rend. Accad. Naz. Lincei* **2015**, *26*, 283–288. [CrossRef]

38. Wang, T.; Franz, T.E.; Yue, W.; Szilagyi, J.; Zlotnik, V.A.; You, J.; Chen, X.; Shulski, M.D.; Young, A. Feasibility analysis of using inverse modeling for estimating natural groundwaterrecharge from a large-scale soil moisture monitoring network. *J. Hydrol.* **2016**, *533*, 250–265. [CrossRef]

39. Bittelli, M. Measuring soil water content: A review. *HortTechnology* **2011**, *21*, 293–300. [CrossRef]

40. Zhang, D.; Tang, R.; Zhao, W.; Tang, B.; Wu, H.; Shao, K.; Li, Z.L. Surface Soil Water Content Estimation from Thermal Remote Sensing based on the Temporal Variation of Land Surface Temperature. *Remote Sens.* **2014**, *6*, 3170–3187. [CrossRef]

41. Zhu, Y.; Wang, Y.; Shao, M.; Horton, R. Estimating soil water content from surface digital image gray level measurements under visible spectrum. *Can. J. Soil Sci.* **2011**, *91*, 69–76. [CrossRef]

42. Kerr, Y.H.; Waldteufel, P.; Wigneron, J.P.; Delwart, S.; Cabot, F.; Boutin, J.; Escorihuela, M.J.; Font, J.; Reul, N.; Gruhier, C.; et al. The SMOS mission: New tool for monitoring key elements of the global water cycle. *Proc. IEEE Inst. Electr. Electron. Eng.* **2010**, *98*, 666–687. [CrossRef]

43. Qiu, B.; Xue, Y.; Fisher, J.B.; Guo, W.; Berry, J.A.; Zhang, Y. Satellite chlorophyll fluorescence and soil moisture observations lead to advances in the predictive understanding of global terrestrial coupled carbon-water cycles. *Glob. Biogeochem. Cycles* **2018**, *32*, 360–375. [CrossRef]

44. Moran, M.S.; Peters–Lidard, C.D.; Watts, J.M.; McElroy, J. Estimating soil moisture at the watershed scale with satellite–based radar and land surface models. *Can. J. Remote Sens.* **2004**, *30*, 805–826. [CrossRef]

45. Hassan, Q.K.; Bourque, C.P.; Meng, F.R.; Cox, R.M. A wetness index using terrain–corrected surface temperature and normalized difference vegetation index derived from standard MODIS products: An evaluation of its use in a humid forest–dominated region of eastern Canada. *Sensors* **2007**, *7*, 2028–2048. [CrossRef]

46. Vitale, M.; Proietti, C.; Cionni, I.; Fischer, R.; De Marco, A. Random forests analysis: A useful tool for defining the relative importance of environmental conditions on crown defoliation. *Water Air Soil Pollut.* **2014**, *225*, 1–17. [CrossRef]

47. Michel, A.; Seidling, W. (Eds.) *Forest Condition in Europe: 2017 Technical Report of ICP Forests. Report under the UNECE Convention on Long-Range Transboundary Air Pollution (CLRTAP)*; BFW Dokumentation 24/2017; BFW Austrian Research Centre for Forests: Vienna, Austria, 2017; 128p.

48. Rossini, M.; Panigada, C.; Meroni, M.; Colombo, R. Assessment of oak forest condition based on leaf biochemical variables and chlorophyll fluorescence. *Tree Physiol.* **2006**, *26*, 1487–1496. [CrossRef] [PubMed]

49. Fischer, R.; Lorenz, M. *Forest Condition in Europe, 2011 Technical Report of ICP Forests and FutMon. Work Report of the Institute for World Forestry 2011/1*; ICP Forests: Hamburg, Germany, 2011; p. 212.

50. De Marco, A.; Proietti, C.; Cionni, I.; Fischer, R.; Screpanti, A.; Vitale, M. Future impacts of nitrogen deposition and climate change scenarios on forest crown defoliation. *Environ. Pollut.* **2014**, *194*, 171–180. [CrossRef]

51. Ferretti, M.; Marchetto, A.; Arisci, S.; Bussotti, F.; Calderisi, M.; Carnicelli, S.; Cecchini, G.; Fabbio, G.; Bertini, G.; Matteucci, G.; et al. On the tracks of Nitrogen deposition effects on temperate forests at their southern European range—An observational study from Italy. *Glob. Chang. Biol.* **2014**, *20*, 3423–3438. [CrossRef]

52. Badea, O. *Manual on Methodology for Long-Term Monitoring of Forest Ecosystems Status under Air Pollution and Climate Change Influence ed*; Editura Silvică: Bucharest, Romania, 2008.

53. ICP Forests. *International Cooperative Programme on Assessment and Monitoring of Air Pollution Effects on Forests. Manual on Methods and Criteria for Harmonized Sampling, Assessment, Monitoring and Analysis of the Effects of Air Pollution on Forests, Part I (Federal Research Center for Forestry and Forest Products*; ICP Forests: Hamburg, Germany, 2010.

54. Zhao, M.; Heinsch, F.A.; Nemani, R.R.; Running, S.W. Improvements of the MODIS terrestrial gross and net primary production global data set. *Remote Sens. Environ.* **2005**, *95*, 164–176. [CrossRef]

55. Turner, D.P.; Ritts, W.D.; Cohen, W.B.; Gower, S.T.; Running, S.W.; Zhao, M.; Costa, M.H.; Kirschbaum, A.A.; Ham, J.M.; Saleska, S.R.; et al. Evaluation of MODIS NPP and GPP products across multiple biomes. *Remote Sens Environ.* **2006**, *102*, 282–292. [CrossRef]

56. Running, S.W.; Nemani, R.R.; Heinsch, F.A.; Zhao, M.; Reeves, M.; Hashimoto, H. A continuous satellite–derived measure of global terrestrial primary production. *Bioscience* **2004**, *54*, 547–560. [CrossRef]

57. Heinsch, F.A.; Reeves, M.; Votava, P.; Kang, S.; Milesi, C.; Zhao, M.; Glassy, J.; Jolly, W.M.; Loehman, R.; Bowker, C.F.; et al. GPP and NPP (MOD17A2/A3) Products NASA MODIS Land Algorithm. In *MOD17 User's Guide*; MODIS Land Team: Washington, DC, USA, 2003; pp. 1–57.

58. Reed, B.C.; Brown, J.F.; VanderZee, D.; Loveland, T.R.; Merchant, J.W.; Ohlen, D.O. Measuring phenological variability from satellite imagery. *J. Veg. Sci.* **1994**, *5*, 703–714. [CrossRef]

59. Stevens, J. Partial and Semipartial Correlations. 2003. Available online: www.uoregon.edu/~{}stevensj/MRA/partial.pdf (accessed on 2010 February 10).

60. Vargas, R.; Baldocchi, D.D.; Bahn, M.; Hanson, P.J.; Hosman, K.P.; Kulmala, L.; Pumpanen, J.; Yang, B. On the multi–temporal correlation between photosynthesis and soil CO_2 efflux: Reconciling lags and observations. *New Phytol.* **2011**, *191*, 1006–1017. [CrossRef]

61. Heinemeyer, A.; Wilkinson, M.; Vargas, R.; Subke, J.A.; Casella, E.; Morison, J.I.L.; Ineson, P. Exploring the "overflow tap" theory: Linking forest soil CO_2 fluxes and individual mycorrhizosphere components to photosynthesis. *Biogeosciences* **2012**, *9*, 79–95. [CrossRef]

62. Proietti, C.; Anav, A.; De Marco, A.; Sicard, P.; Vitale, M. A multi–sites analysis on the ozone effects on Gross Primary Production of European forests. *Sci. Total Environ.* **2016**, *556*, 1–11. [CrossRef]

63. Fares, S.; Vargas, R.; Detto, M.; Goldstein, A.H.; Karlik, J.; Paoletti, E.; Vitale, M. Tropospheric ozone reduces carbon assimilation in trees: Estimates from analysis of continuous flux measurements. *Glob. Chang. Biol.* **2013**, *19*, 2427–2443. [CrossRef] [PubMed]

64. Ceccarelli, T.; Bajocco, S.; Perini, L.; Salvati, L. Urbanisation and Land Take of High Quality Agricultural Soils—Exploring Long–term Land Use Changes and Land Capability in Northern Italy. *Int. J. Environ. Res.* **2014**, *8*, 181–192.

65. Pili, S.; Grigoriadis, E.; Carlucci, M.; Clemente, M.; Salvati, L. Towards Sustainable Growth? A Multi–criteria Assessment of (Changing) Urban Forms. *Ecol. Indic.* **2017**, *76*, 71–80. [CrossRef]

66. Chen, P.Y.; Popovich, P.M. *Correlation: Parametric and Nonparametric Measures*; Sage Publications: Thousand Oaks, CA, USA, 2002.

67. Duvernoy, I.; Zambon, I.; Sateriano, A.; Salvati, L. Pictures from the Other Side of the Fringe: Urban Growth and Peri–urban Agriculture in a Post–industrial City (Toulouse, France). *J. Rural Stud.* **2018**, *57*, 25–35. [CrossRef]

68. Anav, A.; Menut, L.; Khvorostyanov, D.; Viovy, N. Impact of tropospheric ozone on the Euro-Mediterranean vegetation. *Glob. Chang. Biol.* **2011**, *17*, 2342–2359. [CrossRef]

69. Beniston, M. The 2003 heat wave in Europe: A shape of things to come? An analysis based on Swiss climatological data and model simulations. *Geophys. Res. Lett.* **2004**, *31*. [CrossRef]

70. Maracchi, G.; Sirotenko OBindi, M. Impacts of Present and Future Climate Variability on Agriculture and Forestry in the Temperate Regions: Europe. *Clim. Chang.* **2005**, *70*, 117. [CrossRef]

71. Vicente-Serrano, S.M.; Cuadrat-Prats, J.M.; Romo, A. Aridity influence on vegetation patterns in the middle Ebro Valley (Spain): Evaluation by means of AVHRR images and climate interpolation techniques. *J. Arid Environ.* **2006**, *66*, 353–375. [CrossRef]

72. Jump, A.S.; Hunt, J.M.; Peñuelas, J. Rapid climate change–related growth decline at the southern range edge of *Fagus sylvatica*. *Glob. Chang. Biol.* **2006**, *12*, 2163–2174. [CrossRef]

73. Sarris, D.; Christodoulakis, D.; Körner, C. Recent decline in precipitation and tree growth in the eastern Mediterranean. *Glob. Chang. Biol.* **2007**, *13*, 1187–1200. [CrossRef]

74. Martínez-Vilalta, J.; Lopez, B.C.; Adell, N.; Badiella, L.; Ninyerola, M. Twentieth century increase of Scots pine radial growth in NE Spain shows strong climate interactions. *Glob. Chang. Biol.* **2008**, *14*, 2868–2881. [CrossRef]

75. ICP Forests. *International Cooperative Programme on Assessment and Monitoring of Air Pollution Effects on Forests Executive Report*; ICP Forests: Hamburg, Germany, 2008; printed in Germany.

76. De Marco, A.; Sicard, P.; Fares, S.; Tuovinen, J.P.; Anav, A.; Paoletti, E. Assessing the role of soil water limitation in determining the Phytotoxic Ozone Dose (PODY) thresholds. *Atmos. Environ.* **2016**, *147*, 88–97. [CrossRef]

77. Pasho, E.; Camarero, J.J.; de Luis, M.; Vicente-Serrano, S.M. Impacts of drought at different time scales on forest growth across a wide climatic gradient in northeastern Spain. *Agric. For. Meteorol.* **2011**, *151*, 1800–1811. [CrossRef]

78. Araminiene, W.; Sicard, P.; Anav, A.; Agathokleous, E.; Stakėnas, V.; De Marco, A.; Varnagirytė-Kabašinskienė, I.; Paoletti, E.; Girgždienė, R. Trends and inter-relationships of ground-level ozone metrics and forest health in Lithuania. *Sci. Total Environ.* **2019**, *658*, 1265–1277. [CrossRef]

79. Carnicer, J.; Coll, M.; Ninyerola, M.; Pons, X.; Sanchez, G.; Penuelas, J. Widespread crown condition decline, food web disruption, and amplified tree mortality with increased climate change-type drought. *Proc. Natl. Acad. Sci. USA* **2011**, *108*, 1474–1478. [CrossRef] [PubMed]

80. Corcuera, L.; Camarero, J.J.; Gil-Pelegrín, E. Effects of a severe drought on *Quercus ilex* radial growth and xylem anatomy. *Trees-Struct. Funct.* **2004**, *18*, 83–92.

81. Montserrat–Martí, G.; Camarero, J.J.; Palacio, S.; Pérez–Rontomé, C.; Milla, R.; Albuixech, J.; Maestro, M. Summer–drought constrains the phenology and growth of two coexisting Mediterranean oaks with contrasting leaf habit: Implications for their persistence and reproduction. *Trees* **2009**, *23*, 787–799. [CrossRef]

82. Gutiérrez, E.; Campelo, F.; Camarero, J.J.; Ribas, M.; Muntán, E.; Nabais, C.; Freitas, H. Climate controls act at different scales on the seasonal pattern of *Quercus ilex* L. stem radial increments in NE Spain. *Trees Struct. Funct.* **2011**. [CrossRef]

83. Ciais, P.; Reichstein, M.; Viovy, N.; Granier, A.; Ogée, J.; Allard, V.; Aubinet, M.; Buchmann, N.; Bernhofer, C.; Carrara, A.; et al. Europe–wide reduction in primary productivity caused by the heat and drought in 2003. *Nature* **2005**, *437*, 529–533. [CrossRef] [PubMed]

84. Canadell, J.G.; Le Quéré, C.; Raupach, M.R.; Field, C.B.; Buitenhuis, E.T.; Ciais, P.; Conway, T.J.; Gillett, N.P.; Houghton, R.A.; Marland, G. Contributions to accelerating atmospheric CO_2 growth from economic activity, carbon intensity, and efficiency of natural sinks. *Proc. Natl. Acad. Sci. USA* **2007**, *104*, 18866–18870. [CrossRef] [PubMed]

85. Atkinson, M.D. *Betula pendula* Roth (*B. verrucosa* Ehrh.) and *B. pubescens* Ehrh. *J. Ecol.* **1992**, *80*, 837–870. [CrossRef]

86. Bernetti, G. *Selvicoltura Speciale*; Unione Tipografico-Editrice Torinese: Torino, Italy, 1995; ISBN 8802048673.

87. Lebourgeois, F.; Breda, N.; Ulrich, E.; Granier, A. Climate-tree-growth relationships of European beech (*Fagus sylvatica* L.) in the French Permanent Plot Network (RENECOFOR). *Trees* **2005**, *19*, 385–401. [CrossRef]

88. Zhao, M.; Running, S.W. Drought–induced reduction in global terrestrial net primary production from 2000 through 2009. *Science* **2010**, *329*, 940–943. [CrossRef]

89. Churkina, G.; Running, S.W. Contrasting climatic controls on the estimated productivity of global terrestrial biomes. *Ecosystems* **1998**, *1*, 206–215. [CrossRef]

90. Linares, J.C.; Delgado–Huertas, A.; Carreira, J.A. Climatic trends and different drought adaptive capacity and vulnerability in a mixed *Abies pinsapo–Pinus halepensis* forest. *Clim. Chang.* **2010**, *105*, 67–90. [CrossRef]

91. Gruber, A.; Strobl, S.; Veit, B.; Oberhuber, W. Impact of drought on the temporal dynamics of wood formation in *Pinus sylvestris*. *Tree Physiol.* **2010**, *30*, 490–501. [CrossRef] [PubMed]

92. Koepke, D.F.; Kolb, T.E.; Adams, H.D. Variation in woody plant mortality and dieback from severe drought among soils, plant groups, and species within a northern Arizona ecotone. *Glob. Chang. Ecol.* **2010**, *163*, 1079–1090. [CrossRef] [PubMed]

93. Boisvenue, C.; Running, S.W. Impacts of climate change on natural forest productivity—Evidence since the middle of the 20th century. *Glob. Chang. Biol.* **2006**, *12*, 862–882. [CrossRef]

94. Bonan, G.B. Forests and Climate Change: Forcings, Feedbacks, and the Climate Benefits of Forests. *Science* **2008**, *320*, 1444–1449. [CrossRef]
95. Sperry, J.S.; Hacke, U.G.; Oren, R.; Comstock, J.P. Water deficits and hydraulic limits to leaf water supply. *Plant Cell Environ.* **2002**, *25*, 251–263. [CrossRef] [PubMed]

Article

Light Energy Partitioning under Various Environmental Stresses Combined with Elevated CO$_2$ in Three Deciduous Broadleaf Tree Species in Japan

Mitsutoshi Kitao [1,*], Hiroyuki Tobita [2], Satoshi Kitaoka [2], Hisanori Harayama [1], Kenichi Yazaki [2], Masabumi Komatsu [2], Evgenios Agathokleous [3] and Takayoshi Koike [4]

[1] Hokkaido Research Center, Forestry and Forest Products Research Institute, Hitsujigaoka 7, Sapporo 062-8516, Japan; harahisa@ffpri.affrc.go.jp

[2] Department of Plant Ecology, Forestry and Forest Products Research Institute, Matsunosato 1, Tsukuba 305-8687, Japan; tobi@ffpri.affrc.go.jp (H.T.); skitaoka3104@gmail.com (S.K.); kyazaki@ffpri.affrc.go.jp (K.Y.); kopine@ffpri.affrc.go.jp (M.K.)

[3] Institute of Ecology, Key Laboratory of Agrometeorology of Jiangsu Province, School of Applied Meteorology, Nanjing University of Information Science & Technology, Nanjing 210044, China; evgenios@nuist.edu.cn

[4] Laboratory of Plant Nutrition, Hokkaido University, Sapporo 060-8589, Japan; tkoike@for.agr.hokudai.ac.jp

* Correspondence: kitao@ffpri.affrc.go.jp

Received: 28 March 2019; Accepted: 30 May 2019; Published: 3 June 2019

Abstract: Understanding plant response to excessive light energy not consumed by photosynthesis under various environmental stresses, would be important for maintaining biosphere sustainability. Based on previous studies regarding nitrogen (N) limitation, drought in Japanese white birch (*Betula platyphylla* var. *japonica*), and elevated O$_3$ in Japanese oak (*Quercus mongolica* var. *crispula*) and Konara oak (*Q. serrata*) under future-coming elevated CO$_2$ concentrations, we newly analyze the fate of absorbed light energy by a leaf, partitioning into photochemical processes, including photosynthesis, photorespiration and regulated and non-regulated, non-photochemical quenchings. No significant increases in the rate of non-regulated non-photochemical quenching (J$_{NO}$) were observed in plants grown under N limitation, drought and elevated O$_3$ in ambient or elevated CO$_2$. This suggests that the risk of photodamage caused by excessive light energy was not increased by environmental stresses reducing photosynthesis, irrespective of CO$_2$ concentrations. The rate of regulated non-photochemical quenching (J$_{NPQ}$), which contributes to regulating photoprotective thermal dissipation, could well compensate decreases in the photosynthetic electron transport rate through photosystem II (J$_{PSII}$) under various environmental stresses, since J$_{NPQ}$+J$_{PSII}$ was constant across the treatment combinations. It is noteworthy that even decreases in J$_{NO}$ were observed under N limitation and elevated O$_3$, irrespective of CO$_2$ conditions, which may denote a preconditioning-mode *adaptive response* for protection against further stress. Such an *adaptive response* may not fully compensate for the negative effects of lethal stress, but may be critical for coping with non-lethal stress and regulating homeostasis. Regarding the three deciduous broadleaf tree species, elevated CO$_2$ appears not to influence the plant responses to environmental stresses from the viewpoint of susceptibility to photodamage.

Keywords: chlorophyll fluorescence; drought; elevated O$_3$; N limitation; non-photochemical quenching; photodamage

---

1. Introduction

Although light is essential for plant growth, plants can suffer from excessive light, especially when combined with other environmental stresses. Light energy absorbed by a leaf is mainly consumed by photochemical processes such as electron flow to photosynthesis, photorespiration and alternative pathways [1].

Conversely, absorbed light energy is also dissipated by non-photochemical processes divided into two parts: Constitutive, non-regulatory, non-photochemical quenching, and regulatory light-induced, non-photochemical quenching [2–5]. When photosynthetic electron transport is suppressed under environmental stresses, an increase in the fraction of non-regulatory, non-photochemical quenching suggests that plants cannot fully dissipate excess energy through a regulated process [5–9]. Non-regulated, non-photochemical quenching consists of chlorophyll fluorescence internal conversions and intersystem crossing, which leads to the formation of $^1O_2$ via the triplet state of chlorophyll (^{3}chl*) [10–13]. $^1O_2$ can lead to PSII photodamage directly, or via inhibiting PSII repair processes [14–16]. Non-regulated, non-photochemical quenching can be a measure of oxidative stress, as the level of lipid peroxidation indicated by malondialdehyde (MDA) accumulation was closely correlated with the quantum yield of non-regulated, non-photochemical quenching (Y(NO)) in *Arabidopsis thaliana* under a water deficit imposed by withholding the water supply [17].

Plants can acclimate to various environmental conditions by adjusting their leaf physiological characteristics to prevent photodamage [18,19]. For example, within a canopy, sun leaves grown under higher irradiance have a higher photosynthetic capacity with higher area-based leaf nitrogen (N_{area}) than shade leaves [20], a mechanism contributing to maximize photosynthetic carbon gain at the whole plant level by utilizing limited nitrogen optimally [21]. The net photosynthetic rate is known to be proportional to N_{area}, since an increase in N_{area} suggests an increase in Rubisco, a major photosynthetic enzyme [22]. As Rubisco is a key enzyme catalyzing both photosynthesis and photorespiration, electron flow through PSII consumed by the processes also increases with increasing N_{area} [23]. As energy dissipation through photosynthetic electron transport is closely related to N_{area} [23], such N_{area}-related photosynthetic acclimation along the light gradient within the canopy, can also contribute to suppress the risk of photodamage in response to the maximum irradiance during sunflecks, in combination with xanthophyll-related photoprotective energy dissipation [24,25].

Environmental stresses such as N limitation, drought and elevated O_3, causing a reduction in photosynthesis, would increase excessive light energy via a reduction in photosynthetic electron consumption, since photosynthetic carbon assimilation needs NADPH, generated via electron transport [26]. In the coming future, environmental stresses such as nitrogen limitation, as a relative constraint on plant-growth enhancement under elevated CO_2 [27], drought [28], and high O_3 exposure [29–31], are predicted to occur more frequently under global warming and elevated CO_2 concentrations.

We hypothesized that, even under various environmental stresses such as N limitation, drought and elevated O_3 under CO_2 enrichment in the coming future, the non-regulated, non-photochemical quenching should be kept under a certain level, to prevent photosynthetic apparatus from oxidative damage [14–16]. This is achieved by a functional coordination of energy dissipation primarily through N-required electron transport [23], and complementarily through xanthophyll-related thermal energy dissipation, which does not require N investment [8].

To test the hypothesis, we newly analyzed data from previously published works from our research group using three deciduous broadleaf tree species [32–34], where the response of plants to different environmental factors was assessed by chlorophyll fluorescence, so as to assess the fate of absorbed light energy consumed by photochemical processes, and dissipated through constitutively non-regulatory and regulatory light-induced, non-photochemical quenching. We also assessed the light energy not absorbed by a leaf, involved in a bulk loss in chlorophyll pigments, which also has a protective role against photodamage via a reduction in absorbed light energy.

2. Materials and Methods

This study is in part a collective re-analysis of previously published data [32–34]. Regarding "N limitation under elevated CO_2" and "drought under elevated CO_2", regulated and non-regulated non-photochemical quenchings were newly calculated based on data from the previous studies [32,33]. Conversely, regarding "elevated O_3 under elevated CO_2", all data except for A_n were not published previously (cf. [34]).

2.1. N Limitation under Elevated CO_2

Data of Japanese white birch (*Betula platyphylla* var. *japonica*) seedlings grown under limited N and elevated CO_2 were obtained from the study by Kitao et al. [32]. Experiments of N limitation under elevated CO_2 were conducted using a natural daylight phytotron (26/16 °C, day/night; ca. 90% of full sunlight) in Hokkaido Research Center, Forestry and Forest Products Research Institute (FFPRI) in Sapporo, Japan (43°N, 141°E; 180 m above sea level). Details are described in Kitao et al. [32]. One-year-old seedlings of Japanese white birch (*Betula platyphylla* var. *japonica*), a pioneer tree species, 15 to 20 cm in height, were transplanted in free-draining plastic pots filled with clay loam soil mixed with Kanuma pumice soil (1:1 in volume). Pots were placed on trays to prevent nutrient drainage. Each of two CO_2 treatments: 360 μmol mol^{-1} (ambient CO_2 treatment, A-CO_2); and 720 μmol mol^{-1} (elevated CO_2 treatment, E-CO_2) were replicated in two chambers. Two nitrogen levels were applied: 700 mg per plant (adequate nitrogen, +N), or 100 mg per plant (limited nitrogen, -N). The former treatment was conducted as 100 mg N pot^{-1} week^{-1} for 7 weeks during CO_2 treatment, whereas the latter one was conducted as 100 mg N pot^{-1} only once at the onset of CO_2 treatment. We supplied relatively high N for +N treatment to provide adequate N to plants, so as to reach their normal state relative to nursery-grown seedlings. Area-based leaf N (N_{area}) in the seedlings grown in +N treatment was comparable to those grown in the nursery of FFPRI (data not shown). Conversely, we supplied substantially low N for -N treatment, expecting photosynthetic down-regulation under N limitation [35].

2.2. Drought under Elevated CO_2

Data of Japanese white birch seedlings grown under limited water supply and elevated CO_2 were obtained from the study by Kitao et al. [33]. Experiments of drought under elevated CO_2 were also conducted for 1-year-old seedlings of Japanese white birch in the phytotron in Hokkaido Research Center, FFPRI, as described above. Details are described in Kitao et al. [33]. Each of the two CO_2 treatments i.e., 360 (ambient CO_2 treatment: A-CO_2) and 720 μmol mol^{-1} (elevated CO_2 treatment: E-CO_2) was replicated in three chambers. Six randomly selected seedlings in each chamber were supplied daily with 100 mL of water or nutrient solution (once per week) (adequate water supply), while the other six seedlings (totally 12 seedlings) received only 100 mL of nutrient solution once weekly (drought). Each plant received a total of 100 mg N during the experiment, which corresponded to limited N treatment, as described above. The lowest predawn leaf water potential (i.e., measured just prior to the scheduled watering), which was in equilibrium with the soil water potential, was A-CO_2 + adequate water supply: −0.13, A-CO_2 + drought: −0.52, E-CO_2 + adequate water supply: −0.12 and E-CO_2 + drought: −0.39 MPa [33]. The values of water potential in the drought treatment were moderate, since no wilting in the seedlings was observed. Leaves flushed and developed during the drought treatment were used for the measurements.

2.3. Elevated O_3 under Elevated CO_2

Data of Japanese oak (*Quercus mongolica* Fisch. ex Ledeb. var. *crispula* (Blume) H. Ohashi) and Konara oak (*Q. serrata* Murray) seedlings grown under elevated O_3 and CO_2 were obtained from the study by Kitao et al. [34]. Experiments of elevated O_3 under elevated CO_2 were conducted in a free-air concentration-enrichment (FACE) exposure system, consisting of 12 plots (3 replicates per treatment), located at the nursery of FFPRI in Tsukuba, Japan (36°00′N, 140°08′E, 20 m a.s.l.).

Details are described in Kitao et al. [34]. One-year-old seedlings of Japanese oak and Konara oak, gap-dependent mid-successional tree species, approximately 5 cm in height under dormancy, were transplanted directly to the ground in the plots. The treatments were as follows: Control (unchanged ambient air), elevated CO_2 (Target set, 550 μmol mol^{-1}), elevated O_3 (Target set, twice-ambient), and elevated $CO_2 + O_3$ (550 µmol mol^{-1} CO_2 and twice-ambient O_3). Plants were grown under the treatments for two growing seasons. Measurements of gas exchange and chlorophyll fluorescence were conducted in the second growing season.

2.4. Measurements of Gas Exchange and Chlorophyll Fluorescence

Measurements of gas exchange and chlorophyll fluorescence were conducted with a portable photosynthesis measuring system (Li-6400, Li-Cor, Lincoln, NE, USA), combined with a portable fluorometer (PAM-2000, Walz, Effeltrich, Germany) for plants grown under "N limitation with CO_2 enrichment", or a leaf chamber fluorometer (Li-6400-40, Li-Cor) for plants grown under "drought, and elevated O_3 under elevated CO_2". Details are described in Kitao et al. [32–34]. The net photosynthetic rate (A_n), quantum yield of PSII electron transport (Y(II)), quantum yield of non-regulate, non-photochemical quenching in PSII (Y(NO)), and finally the quantum yield of regulated, non-photochemical quenching in PSII (Y(NPQ)) [2–5] were measured at a photosynthetic steady state under saturating light intensities provided by a red/blue LED array (Li-6400-40, Li-Cor), with blue light comprising 10% of the total PPFD. We measured Y(NO) and Y(NPQ), based on the simple approach: Y(NO) = F/F_m, and Y(NPQ) = $F/F_m' - F/F_m$, where F, F_m and F_m' is the relative fluorescence yield at steady state illumination, the relative maximum fluorescence yield in dark-adapted conditions, or that during illumination, respectively [4,5]. Regarding the data sets of drought and elevated O_3 under elevated CO_2, we measured F, F_m' and F_o' (the minimum fluorescence yield during illumination) during the gas exchange measurements, but did not measure F_m. We measured F_v/F_m on the following day, after an overnight dark-adaptation in the same leaves for the gas exchange, and chlorophyll fluorescence measurements with the photosynthesis system (Li-6400, Li-Cor) for drought-treated plants [33], and with a portable fluorometer (Mini-PAM, Walz) for O_3-treated plants [34]. Since F_o' is estimated as $F_o' = F_o/(F_v/F_m + F_o/F_m')$ [36], F_m can be estimated as $F_m = (F_m' \times F_o' \times F_v/F_m)/((F_m' - F_o') \times (1 - F_v/F_m))$. This would be a practical approach to determine F_m for many samples in the field after the fluorescence measurements during daytime. Leaf absorptance (ABS) was calculated from a calibration curve between SPAD readings (measured with a SPAD chlorophyll meter, SPAD 502, Minolta, Osaka, Japan) and leaf absorptance [32–34]. Based on the chlorophyll fluorescence parameters, the electron transport rate (J_{PSII}) was calculated as J_{PSII} = Y(II) × ABS × light intensity × 0.5 [6]. Analogous to J_{PSII}, the rate of regulatory thermal dissipation (J_{NPQ}) and the rate of non-regulatory energy dissipation via heat or fluorescence (J_{NO}) were estimated from Y(NPQ) × ABS × light intensity × 0.5 and Y(NO) × ABS × light intensity × 0.5, respectively [4]. Light energy not absorbed by chlorophyll in a leaf (J_{Chl}) was estimated as J_{Chl} = (1 − ABS) × light intensity × 0.5.

2.5. Leaf N Content

Regarding '*elevated O_3 under elevated CO_2*', the leaves were sampled after the measurements and used for a determination of N_{area} by the combustion method, using an analysis system composed of an N/C determination unit (SUMIGRAPH, NC 800, Sumika Chem. Anal. Service, Osaka, Japan), a gas chromatograph (GC 8A, Shimadzu, Kyoto, Japan), and a data processor (Chromatopac, C R6A, Shimadzu).

2.6. Statistical Analysis

In the study on N limitation under elevated CO_2, individual seedlings across the two chambers were used as the sample unit (n = 4–6). Two-way Analysis of Variance (ANOVA) (N × CO_2) was used to test the differences in the treatment means of A_n, J_{PSII}, J_{NPQ}, $J_{PSII}+J_{NPQ}$, J_{NO} and J_{Chl}. In the study on drought under elevated CO_2, statistics are based on the individual plot (CO_2 × water regime) in each chamber as the sample unit (n = 3). Three to six plants were measured in each plot.

A mean value from these plants was used as the estimate for that sample unit. Two-way ANOVA, with one between-subjects factor (CO_2) and one within-subject factor (water regime), was used to test treatment differences in A_n, J_{PSII}, J_{NPQ}, $J_{PSII} + J_{NPQ}$, J_{NO} and J_{Chl}. In the study on elevated O_3 under elevated CO_2, all statistics were based on the mean value of the individual plot ($CO_2 \times O_3$ regime) as the sample unit ($n = 3$). These values were then averaged to provide the sample estimate for that replicate. Three-way ANOVA, with two between-subjects factors (CO_2 and O_3) and one within-subject factor (species), was used to test the differences in A_n, J_{PSII}, J_{NPQ}, $J_{PSII} + J_{NPQ}$, J_{NO} and J_{Chl}, and leaf N.

3. Results

3.1. Nitrogen Limitation under Elevated CO_2

When compared at the growth CO_2, i.e., 360 µmol mol^{-1} for the ambient-CO_2-grown plants, and 720 µmol mol^{-1} for the elevated-CO_2-grown plants, higher A_n was observed in plants grown under elevated CO_2 than in ambient-CO_2 plants with adequate N supply, whereas no enhancement in A_n under elevated CO_2 was observed with a limited N supply (Figure 1, Table 1). Conversely, no enhancement in J_{PSII} was observed in plants grown under elevated CO_2 with an adequate N supply, whereas the limited N supply resulted in lower J_{PSII} irrespective of CO_2 treatments. J_{NPQ} was significantly higher in plants grown under limited N supply than those under adequate N supply. The sum of $J_{NPQ} + J_{PSII}$ was not significantly different among the treatment combinations. As ABS was lower in the plants grown with limited N supply, higher J_{Chl} was observed in those plants. As a consequence of the increased J_{Chl} in addition to J_{NPQ}, lower J_{NO} was observed in the plants grown with limited N supply, in spite of significantly lower J_{PSII}, irrespective of CO_2 treatment.

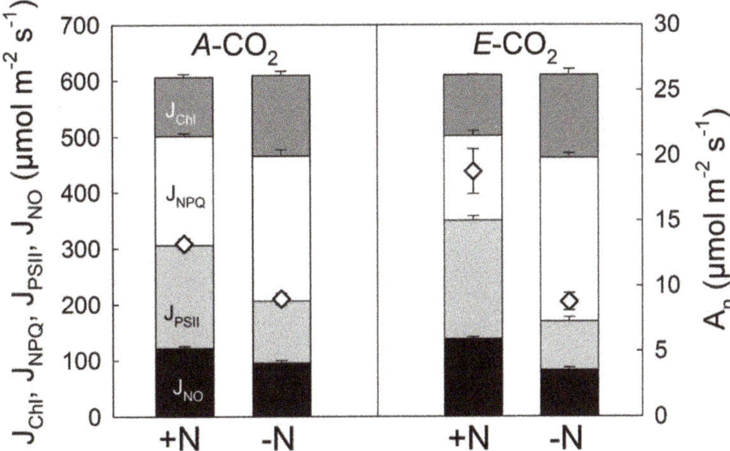

Figure 1. Fate of light energy partitioning in the seedlings of Japanese white birch grown with N limitation under elevated CO_2. J_{Chl}, J_{NPQ}, J_{PSII} and J_{NO} were measured in the seedlings of Japanese white birch grown under ambient (A-CO_2) and elevated CO_2 (E-CO_2) with adequate (+N) and limited N supply (-N). Open diamonds indicate net photosynthetic rate (A_n). Measurements were conducted for fully-developed mature leaves (leaf age was approx. 1 month) under respective growth CO_2 concentrations (i.e., 360 µmol mol^{-1} for ambient-CO_2-grown plants, and 720 µmol mol^{-1} for elevated-CO_2-grown plants) at saturating light (1200 µmol m^{-2} s^{-1}). Values are means ± se ($n = 4$–6). Data were obtained from Kitao et al. [32]. J_{Chl}, J_{NPQ} and J_{NO} were newly calculated based on data from the previous study.

Table 1. F values of Analysis of Variance (ANOVA) to test the effects of various environmental stresses (N limitation, drought and elevated O_3) under ambient or elevated CO_2 on J_{NO}, J_{PSII}, J_{NPQ}, $J_{PSII}+J_{NPQ}$, J_{Chl}, and A_n measured at respective growth CO_2 concentrations. Significant effects are indicated in the table by ***: $p \leq 0.001$, **: $p \leq 0.01$, *: $p \leq 0.05$, and ns: non-significant. Data were obtained from Kitao et al. [32–34].

Treatment	Effect	J_{NO}	J_{PSII}	J_{NPQ}	$J_{PSII} + J_{NPQ}$	J_{Chl}	A_n
				F-statistics			
N limitation	CO_2 ($F_{1,13}$)	0.69 ns	0.47 ns	0.81 ns	0.11 ns	0.03 ns	51.8 ***
	N ($F_{1,13}$)	106 ***	133 ***	124 ***	0.19 ns	28.6 ***	322 ***
	$CO_2 \times$ N ($F_{1,13}$)	13.3 **	9.63 **	18.4 ***	3.08 ns	0.00 ns	71.4 ***
Drought	CO_2 ($F_{1,4}$)	0.01 ns	25.1 **	2.68 ns	0.00 ns	0.01 ns	3.75 ns
	Drought ($F_{1,4}$)	0.71 ns	0.40 ns	3.90 ns	2.23 ns	1.39 ns	0.74 ns
	$CO_2 \times$ Drought ($F_{1,4}$)	0.02 ns	0.40 ns	1.86 ns	1.28 ns	2.04 ns	0.29 ns
Elevated O_3	CO_2 ($F_{1,8}$)	5.00 ns	5.46 *	16.4 **	0.50 ns	0.05 ns	35.6 ***
	O_3 ($F_{1,8}$)	5.54 *	6.60 *	14.9 **	0.07 ns	0.72 ns	25.5 ***
	$CO_2 \times O_3$ ($F_{1,8}$)	0.91 ns	1.35 ns	0.08 ns	2.54 ns	2.46 ns	0.94 ns
	Species ($F_{1,8}$)	1.36 ns	12.7 ***	0.42 ns	4.36 ns	2.13 ns	39.3 ***
	$CO_2 \times$ Species ($F_{1,8}$)	0.08 ns	2.35 ns	0.22 ns	0.50 ns	0.35 ns	6.64 *
	$O_3 \times$ Species ($F_{1,8}$)	0.10 ns	18.0 **	1.15 ns	4.79 ns	4.32 ns	8.59 *
	$CO_2 \times O_3 \times$ Species ($F_{1,8}$)	4.43 ns	23.9 **	7.58 *	0.97 ns	6.32 *	2.30 ns

3.2. Drought under Elevated CO_2

Measurements of gas exchange and chlorophyll fluorescence were conducted at the growth CO_2 (i.e., 360 µmol mol^{-1} for the ambient-CO_2-grown plants and 720 µmol mol^{-1} for the elevated-CO_2-grown plants) when soils were most dried on the previous day of irrigation (i.e., just prior to the scheduled watering). Intercellular CO_2 concentration (C_i) was higher under elevated CO_2, but lower under drought (Figure 2). Irrespective of the large variation of C_i itself, J_{Chl}, J_{NPQ}, J_{NO} and $J_{PSII}+J_{NPQ}$ were not significantly different among the treatment combinations (Figure 2, Table 1). Only J_{PSII} was significantly lower in the plants grown under elevated CO_2, whereas no significant difference in A_n was observed among the treatment combinations.

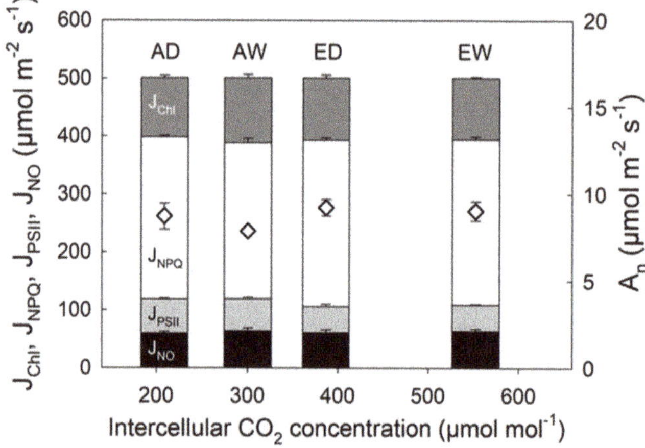

Figure 2. Fate of light energy partitioning with drought under elevated CO_2. Data are plotted as a function of intercellular CO_2 concentration. J_{Chl}, J_{NPQ}, J_{PSII} and J_{NO} were measured in the seedlings

of Japanese white birch grown under ambient and elevated CO_2 with adequate (daily) and limited (once-weekly) water supply. AD: Ambient CO_2 + once-weekly irrigation; AW: Ambient CO_2 + daily-irrigation; ED: Elevated CO_2 + once-weekly irrigation; EW: Elevated CO_2 + daily-irrigation. Open diamonds indicate net photosynthetic rate (A_n). Measurements were conducted for fully-developed mature leaves (leaf age was approx. 1 month) under the most dried conditions under respective growth CO_2 concentrations (i.e., 360 µmol mol^{-1} for ambient-CO_2-grown plants, and 720 µmol mol^{-1} for elevated-CO_2-grown plants) at saturating light (1000 µmol m^{-2} s^{-1}). Values are means ± se (n = 3). Data were obtained from Kitao et al. [33]. J_{Chl}, J_{NPQ} and J_{NO} were newly calculated based on data from the previous study.

3.3. Elevated O_3 under Elevated CO_2

A_n measured at the respective growth CO_2 (i.e., 380 µmol mol^{-1} for the ambient-CO_2-grown plants and 550 µmol mol^{-1} for the elevated-CO_2-grown plants) increased under elevated CO_2, but decreased under elevated O_3 (Figure 3, Table 1). A_n was significantly different between *Q. mongolica* and *Q. serrata*, and the effects of CO_2 and O_3 were also different between species (Table 1). J_{PSII} increased under elevated CO_2, but decreased under elevated O_3, whereas J_{NPQ} decreased under elevated CO_2 but increased under elevated O_3. As a result, no significant differences were observed in $J_{PSII}+J_{NPQ}$ among the treatment combinations or across species. J_{Chl} was neither affected by CO_2, O_3 nor species. Significantly lower J_{NO} was observed in the plants grown under elevated O_3. Area-based leaf N content (N_{area}) was not significantly different among the treatment combinations, whereas significantly higher N_{area} was observed in *Q. serrata* (Figure 4, Table 2).

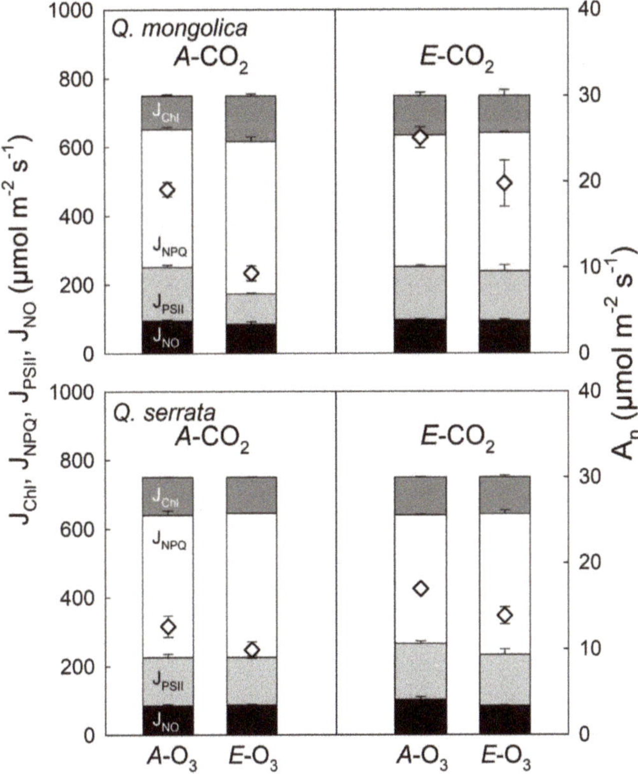

Figure 3. Fate of light energy partitioning under elevated O_3 and CO_2. Here, J_{Chl}, J_{NPQ}, J_{PSII} and J_{NO} were measured in the seedlings of Japanese oak (*Q. mongolica*) and Konara oak (*Q. serrata*) grown

under ambient (*A*-O$_3$) and elevated O$_3$ (*E*-O$_3$), combined with ambient (*A*-CO$_2$) and elevated CO$_2$ (*E*-CO$_2$). Open diamonds indicate the net photosynthetic rate (A$_n$). Measurements were conducted for fully-developed mature leaves (leaf age was approx. 2 months) under respective growth CO$_2$ concentrations (i.e., 380 µmol mol^{-1} for ambient-CO$_2$-grown plants, and 550 µmol mol^{-1} for elevated-CO$_2$-grown plants) at saturating light (1500 µmol m^{-2} s^{-1}). Values are means ± se (*n* = 3). A$_n$ was obtained from Kitao et al. [34].

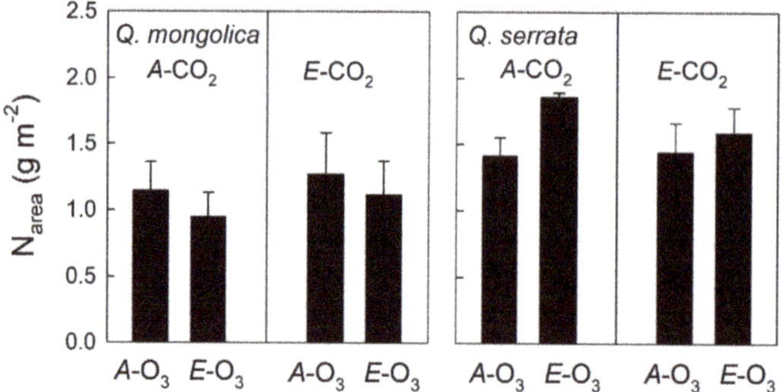

Figure 4. Area-based leaf N content (N$_{arae}$) in the seedlings of Japanese oak (*Q. mongolica*) and Konara oak (*Q. serrata*) grown under ambient (*A*-O$_3$) and elevated O$_3$ (*E*-O$_3$), combined with ambient (*A*-CO$_2$) and elevated CO$_2$ (*E*-CO$_2$). Values are means ± se (*n* = 3).

Table 2. *F* values of three-way ANOVA with two between-subjects factors (CO$_2$ and O$_3$) and one within-subject factor (species), to test the effects of CO$_2$ ($F_{1,8}$), O$_3$ ($F_{1,8}$), species ($F_{1,8}$), CO$_2$ × O$_3$ ($F_{1,8}$), CO$_2$ × Species ($F_{1,8}$), O$_3$ × Species ($F_{1,8}$), and CO$_2$ × O$_3$ × Species ($F_{1,8}$) on area-based leaf nitrogen content (N$_{area}$). The symbols ns and * denote non-significant (*p* > 0.05) and significant (*p* ≤ 0.05) effects, respectively.

	F-Statistics
Effect	**N$_{area}$**
CO$_2$	0.01 [ns]
O$_3$	0.25 [ns]
CO$_2$ × O$_3$	0.27 [ns]
Species	7.03 *
CO$_2$ × Species	0.58 [ns]
O$_3$ × Species	1.94 [ns]
CO$_2$ × O$_3$ × Species	0.25 [ns]

4. Discussion

4.1. Nitrogen Limitation under Elevated CO$_2$

Nitrogen plays a key role in photosynthesis, since Rubisco, a key-enzyme of photosynthesis, is the largest sink of N in a leaf [22], and also a considerable amount of N is involved in proteins related to linear electron transport [23,37]. Plants grown under elevated CO$_2$ often show photosynthetic acclimation typically accompanied with a decrease in the maximum capacity of Rubisco carboxylation, known as photosynthetic down-regulation, particularly under limited nitrogen availability [27,35]. When Japanese white birch seedlings were grown under the combinations of CO$_2$ and N treatments, leaves showed higher N$_{area}$ with higher N supply, but lower N$_{area}$ under elevated CO$_2$ treatment [32]. In the present study, elevated CO$_2$ had no effect on J$_{NO}$, whereas limited N decreased J$_{NO}$, suggesting a lower risk of photodamage under N limitation, irrespective of lower A$_n$ [5]. The decreases in electron transport rate (J_{PSII}) by N limitation and photosynthetic down-regulation under elevated CO_2 were fully-compensated by regulated thermal energy dissipation (J_{NPQ}), since the sum of J_{PSII} and J_{NPQ} was not significantly different across the treatment combinations. Conversely, the decrease in J_{NO} under limited N resulted mainly from the increased loss of absorbed light energy, indicated by the increase in J_{Chl}.

4.2. Drought under Elevated CO$_2$

Drought-induced stomatal closure leads to low intercellular CO_2 (C_i) [1]. Leaves developed under long-term drought display higher photosynthetic capacity, accompanied with higher N_{area}, thus compensating the reduced photosynthetic performance under low C_i [38–40]. In the present study, seedlings of Japanese white birch were grown under elevated CO_2 and long-term drought with limited N supply. Photosynthetic capacity, indicated by the maximum rate of Rubisco carboxylation ($V_{c,max}$), was previously shown to increase by long-term drought accompanied with higher N_{area}, whereas elevated CO_2 decreased $V_{c,max}$ with lower N_{area} [33]. In combination of changes in $V_{c,max}$ with different C_i, A_n was not significantly different among the treatment combinations. In spite of similar A_n, J_{PSII} decreased under elevated CO_2, maybe because of a suppression of photorespiration under elevated CO_2 (720 μmol mol^{-1}) [41]. The decrease in J_{PSII} under elevated CO_2 was well compensated by a regulated photoprotective reaction (J_{NPQ}) [2,5], leading to unchanged J_{NO} under the combinations of CO_2 and water treatments. An increase in Y(NO) was reported in mature leaves of *A. thaliana* under water deficit by withholding water, whereas a less extent of increase in Y(NO) was observed in young leaves, suggesting higher acclimating capacity, preventing oxidative damage in younger leaves [17]. In the present study, as the leaves of Japanese birch seedlings had flushed and developed during the relatively moderate drought treatment, they might fully acclimate to long-term drought, preventing photodamage [38–40].

4.3. Elevated O$_3$ under Elevated CO$_2$

Tropospheric ozone (O_3) levels continue to increase globally [42,43], concurrently occurring with an increase in atmospheric CO_2 concentration [44]. Contrary to elevated CO_2, which may enhance plant growth in the short term [45,46], elevated O_3 generally reduces plant growth via a reduction in photosynthetic rate and increased respiration rate [30,47]. Deciduous broadleaf trees native to Japan, Japanese oak (*Quercus mongolica*) and Konara oak (*Q. serrata*), were exposed to free air enriched with elevated O_3 (twice ambient O_3) and/or CO_2 (550 μmol mol^{-1} as target). A_n in the fully-expanded second-flushed leaves, measured at each growth CO_2, reduced by elevated O_3 but enhanced by elevated CO_2, irrespective of species. As A_n was enhanced under elevated CO_2 with no difference in N_{area} among the treatment combinations, photosynthetic down-regulation, which is often induced by elevated CO_2 under limited N availability [32,35], was not apparent in the present study of a free-air CO_2 and O_3 exposure without limitations of root growth [34]. Furthermore, reduced leaf N, accompanied with a reduction in A_n under elevated O_3 [48], was not observed in the present study, suggesting that causes other than leaf N reduction might be predominant to decrease A_n, such as an oxidative stress in the chloroplast [49]. J_{PSII} was also reduced by elevated O_3, but increased by elevated CO_2, as well as A_n. In contrast, J_{NPQ} was increased by elevated O_3, but decreased by elevated CO_2, which might fully compensate the changes in J_{PSII}, as indicated by the constant $J_{PSII}+J_{NPQ}$. It is noteworthy that J_{NO} decreased under elevated O_3, which means that elevated O_3 would not necessarily increase the risk of photodamage in these species.

4.4. Regulated and Non-regulated Non-photochemical Quenching under Elevated CO$_2$

In the present study, we investigated the fate of light energy absorbed by a leaf under various environmental stresses combined with elevated CO_2. We particularly focused on J_{NO}, a measure of constitutive, non-regulated, non-photochemical energy dissipation, because an increase in J_{NO} suggests an increase in the risk of photodamage [2,5]. As a whole, photoprotective thermal energy dissipation indicated by J_{NPQ} may well compensate for the decreases in J_{PSII} under environmental stresses, since $J_{PSII}+J_{NPQ}$ was rather constant throughout the various stresses, even under elevated CO_2. If plants can keep J_{PSII} constant, there is a high potential for preventing the accumulation of excess energy [25,38–40]. However, if J_{PSII} is restricted under limited N supply or by other environmental stress such as elevated O_3, xanthophyll-related regulated thermal energy dissipation (J_{NPQ}) would act as an efficient safety valve, which does not need N investment [8].

Furthermore, although drought and elevated CO_2 had no effects on J_{NO}, N limitation and elevated O_3 resulted in decreases in J_{NO}, in contrast to expected stress responses (i.e., increases in J_{NO}), which can be considered as an *adaptive response* in the framework of pre-conditioning to cope with further environmental stresses [50,51]. By doing so, J_{NO} may be decreased to such an extent that will offset high increases that would occur under further stress. This novel mechanism builds upon an extended body of literature showing the biological capacity of a variety of organisms to display hormetic *adaptive responses* which eventually act as biological shields against following health threats [50–52]. Such *adaptive responses* for coping with stress are activated by low/mild severity of stress, at levels that are (often far) lower than the level beyond which toxicological, adverse responses occur [50–52]. This suggests that NPQ can compensate for the effects of following more severe environmental stress, but if the stress is too severe (e.g., acute exposure), increased NPQ may not be enough to fully compensate for the negative effects of stress.

Whereas it was difficult to explicitly determine the factor inducing lower J_{NO} under elevated O_3 (maybe the integrated effects of $J_{NPQ} + J_{Chl}$), an increase in J_{Chl} apparently contributed to reducing J_{NO} under limited N. Thus, in addition to the fractions of absorbed light energy partitioning, based on chlorophyll fluorescence (Y(II), Y(NPQ) and Y(NO)), reduced chlorophyll pigments should be taken into account as a photoprotective reaction for assessing environmental stresses by using chlorophyll fluorescence measurements [53].

Similar to the present study, a stable or even lower Y(NO) due to the decline in Y(II), accompanied with the increase in Y(NPQ), was also reported in paraquat-exposed *Arabidopsis thaliana* [11,18] and in Al-exposed *A. thaliana* [12]. The decrease in J_{NO} may denote also decreased ROS production [17]. Non-regulated, non-photochemical quenching consists of chlorophyll fluorescence internal conversions and intersystem crossing, which leads to the formation of singlet oxygen ($^1O_2$) via the triplet state of chlorophyll (^{3}chl*) [10,11,13]. Since J_{NO} declined, it seems that J_{NPQ} was sufficient enough to protect plants from ROS, by exhibiting lower $^1O_2$ production, and preventing the photosynthetic apparatus from oxidative damage [12].

5. Conclusions

Based on the results from three deciduous broadleaf tree species in the present study, even when photosynthesis and J_{PSII} were reduced by environmental stresses, photoprotective mechanisms including J_{NPQ} and J_{Chl} could suppress the rise of J_{NO} in the leaves developed under the stresses, consequently preventing photodamage even under future-coming elevated CO_2 conditions.

Author Contributions: M.K., and H.T. designed the study. M.K., H.T., S.K., H.H., K.Y. and M.K. collected the photosynthetic data, performed the analysis, and hence equally contributed to this study. M.K. led the writing with input from E.A. and T.K. All authors also discussed the results and commented on the manuscript.

Funding: This study was supported in part by JSPS KAKENHI Grant Number JP17K19301, JP17F17102 and JP17H03839. Evgenios Agathokleous was a JSPS International Research Fellow (ID No: P17102).

Acknowledgments: We thank K Mima and K Sakai for leaf N analyses, and express our sincere appreciation to V. Hurry for his helpful suggestions on our article. E.A. acknowledges multi-year support from The Startup Foundation for Introducing Talent of Nanjing University of Information Science & Technology (NUIST), Nanjing, China.

Conflicts of Interest: The authors declare no conflict of interest.

References

1. Cornic, G.; Fresneau, C. Photosynthetic carbon reduction and carbon oxidation cycles are the main electron sinks for photosystem II activity during a mild drought. *Ann. Bot.* **2002**, *89*, 887–894. [CrossRef] [PubMed]
2. Lazár, D. Parameters of photosynthetic energy partitioning. *J. Plant Physiol.* **2015**, *175*, 131–147. [CrossRef]
3. Kramer, D.M.; Johnson, G.; Kiirats, O.; Edwards, G.E. New fluorescence parameters for the determination of Q_A redox state and excitation energy fluxes. *Photosynth. Res.* **2004**, *79*, 209–218. [CrossRef] [PubMed]
4. Hendrickson, L.; Furbank, R.T.; Chow, W.S. A simple alternative approach to assessing the fate of absorbed light energy using chlorophyll fluorescence. *Photosynth. Res.* **2004**, *82*, 73–81. [CrossRef] [PubMed]
5. Klughammer, C.; Schreiber, U. Complementary PS II quantum yields calculated from simple fluorescence parameters measured by PAM fluorometry and the Saturation Pulse method. *PAM Appl. Notes* **2008**, *1*, 27–35.
6. Demmig-Adams, B.; Adams, W.W., III; Barker, D.H.; Logan, B.A.; Bowling, D.R.; Verhoeven, A.S. Using chlorophyll fluorescence to assess the fraction of absorbed light allocated to thermal dissipation of excess excitation. *Physiol. Plant.* **1996**, *98*, 253–264. [CrossRef]
7. Ruban, A.V.; Berera, R.; Ilioaia, C.; van Stokkum, I.H.M.; Kennis, J.T.M.; Pascal, A.A.; van Amerongen, H.; Robert, B.; Horton, P.; van Grondelle, R. Identification of a mechanism of photoprotective energy dissipation in higher plants. *Nature* **2007**, *450*, 575–578. [CrossRef]
8. Wilhelm, C.; Selmar, D. Energy dissipation is an essential mechanism to sustain the viability of plants: The physiological limits of improved photosynthesis. *J. Plant Physiol.* **2011**, *168*, 79–87. [CrossRef]
9. Kornyeyev, D.; Logan, B.A.; Holaday, A.S. Excitation pressure as a measure of the sensitivity of photosystem II to photoinactivation. *Funct. Plant Biol.* **2010**, *37*, 943–951. [CrossRef]
10. Kasajima, I.; Ebana, K.; Yamamoto, T.; Takahara, K.; Yano, M.; Kawai-Yamada, M.; Uchimiya, H. Molecular distinction in genetic regulation of nonphotochemical quenching in rice. *Proc. Natl. Acad. Sci. USA* **2011**, *108*, 13835–13840. [CrossRef]
11. Moustaka, J.; Tanou, G.; Adamakis, I.D.; Eleftheriou, E.P.; Moustakas, M. Leaf age-dependent photoprotective and antioxidative response mechanisms to paraquat-induced oxidative stress in *Arabidopsis thaliana*. *Int. J. Mol. Sci.* **2015**, *16*, 13989–14006. [CrossRef]
12. Moustaka, J.; Ouzounidou, G.; Sperdouli, I.; Moustakas, M. Photosystem II is more sensitive than photosystem I to Al^{3+} induced phytotoxicity. *Materials (Basel)* **2018**, *11*, 1772. [CrossRef] [PubMed]
13. Gawroński, P.; Witoń, D.; Vashutina, K.; Bederska, M.; Betliński, B.; Rusaczonek, A.; Karpiński, S. Mitogen-activated protein kinase 4 is a salicylic acid-independent regulator of growth but not of photosynthesis in *Arabidopsis*. *Mol. Plant* **2014**, *7*, 1151–1166. [CrossRef] [PubMed]
14. Krieger-Liszkay, A.; Fufezan, C.; Trebst, A. Singlet oxygen production in photosystem II and related protection mechanism. *Photosynth. Res.* **2008**, *98*, 551–564. [CrossRef] [PubMed]
15. Vass, I. Molecular mechanisms of photodamage in the Photosystem II complex. *Biochim. Biophys. Acta Bioenerg.* **2012**, *1817*, 209–217. [CrossRef] [PubMed]
16. Vass, I. Role of charge recombination processes in photodamage and photoprotection of the photosystem II complex. *Physiol. Plant.* **2011**, *142*, 6–16. [CrossRef]
17. Sperdouli, I.; Moustakas, M. A better energy allocation of absorbed light in photosystem II and less photooxidative damage contribute to acclimation of *Arabidopsis thaliana* young leaves to water deficit. *J. Plant Physiol.* **2014**, *171*, 587–593. [CrossRef] [PubMed]
18. Moustaka, J.; Moustakas, M. Photoprotective mechanism of the non-target organism *Arabidopsis thaliana* to paraquat exposure. *Pestic. Biochem. Physiol.* **2014**, *111*, 1–6. [CrossRef]
19. Liu, J.; Zhang, R.; Zhang, G.; Guo, J.; Dong, Z. Effects of soil drought on photosynthetic traits and antioxidant enzyme activities in *Hippophae rhamnoides* seedlings. *J. For. Res.* **2017**, *28*, 255–263. [CrossRef]
20. Keenan, T.F.; Niinemets, Ü. Global leaf trait estimates biased due to plasticity in the shade. *Nat. Plants* **2016**, *3*, 16201. [CrossRef] [PubMed]
21. Hikosaka, K.; Anten, N.P.R.; Borjigidai, A.; Kamiyama, C.; Sakai, H.; Hasegawa, T.; Oikawa, S.; Iio, A.; Watanabe, M.; Koike, T.; et al. A meta-analysis of leaf nitrogen distribution within plant canopies. *Ann. Bot.* **2016**, *118*, 239–247. [CrossRef]
22. Evans, J.R. Photosynthesis and nitrogen relationship in leaves of C3 plants. *Oecologia* **1989**, *78*, 9–19. [CrossRef] [PubMed]

23. Niinemets, Ü.; Bilger, W.; Kull, O.; Tenhunen, J. Responses of foliar photosynthetic electron transport, pigment stoichiometry, and stomatal conductance to interacting environmental factors in a mixed species forest canopy. *Tree Physiol.* **1999**, *19*, 839–852. [CrossRef] [PubMed]

24. Kitao, M.; Kitaoka, S.; Komatsu, M.; Utsugi, H.; Tobita, H.; Koike, T.; Maruyama, Y. Leaves of Japanese oak (*Quercus mongolica* var. *crispula*) mitigate photoinhibition by adjusting electron transport capacities and thermal energy dissipation along the intra-canopy light gradient. *Physiol. Plant.* **2012**, *146*, 192–204. [CrossRef] [PubMed]

25. Kitao, M.; Kitaoka, S.; Harayama, H.; Tobita, H.; Agathokleous, E.; Utsugi, H. Canopy nitrogen distribution is optimized to prevent photoinhibition throughout the canopy during sun flecks. *Sci. Rep.* **2018**, *8*, 503. [CrossRef]

26. Von Caemmerer, S. *Biochemical Models of Leaf Photosynthesis*; CSIRO Pub: Clayton, Australia, 2000; ISBN 064306379X.

27. Coskun, D.; Britto, D.T.; Kronzucker, H.J. Nutrient constraints on terrestrial carbon fixation: The role of nitrogen. *J. Plant Physiol.* **2016**, *203*, 95–109. [CrossRef] [PubMed]

28. Dai, A. Increasing drought under global warming in observations and models. *Nat. Clim. Chang.* **2012**, *3*, 52–58. [CrossRef]

29. Ainsworth, E.A.; Yendrek, C.R.; Sitch, S.; Collins, W.J.; Emberson, L.D. The effects of tropospheric ozone on net primary productivity and implications for climate change. *Annu. Rev. Plant Biol.* **2012**, *63*, 637–661. [CrossRef] [PubMed]

30. Matyssek, R.; Wieser, G.; Ceulemans, R.; Rennenberg, H.; Pretzsch, H.; Haberer, K.; Löw, M.; Nunn, A.J.J.; Werner, H.; Wipfler, P.; et al. Enhanced ozone strongly reduces carbon sink strength of adult beech (*Fagus sylvatica*)—Resume from the free-air fumigation study at Kranzberg Forest. *Environ. Pollut.* **2010**, *158*, 2527–2532. [CrossRef] [PubMed]

31. Sharma, C.; Kontunen-Soppela, S.; Pandey, A.K.; Keski-Saari, S.; Oksanen, E.; Pandey, V. Impacts of increasing ozone on Indian plants. *Environ. Pollut.* **2013**, *177*, 189–200.

32. Kitao, M.; Koike, T.; Tobita, H.; Maruyama, Y. Elevated CO_2 and limited nitrogen nutrition can restrict excitation energy dissipation in photosystem II of Japanese white birch (*Betula platyphylla* var. *japonica*) leaves. *Physiol. Plant.* **2005**, *125*, 64–73. [CrossRef]

33. Kitao, M.; Lei, T.T.; Koike, T.; Kayama, M.; Tobita, H.; Maruyama, Y. Interaction of drought and elevated CO_2 concentration on photosynthetic down-regulation and susceptibility to photoinhibition in Japanese white birch seedlings grown with limited N availability. *Tree Physiol.* **2007**, *27*, 727–735. [CrossRef] [PubMed]

34. Kitao, M.; Komatsu, M.; Yazaki, K.; Kitaoka, S.; Tobita, H. Growth overcompensation against O_3 exposure in two Japanese oak species, *Quercus mongolica* var. *crispula* and *Quercus serrata*, grown under elevated CO_2. *Environ. Pollut.* **2015**, *206*, 133–141. [CrossRef] [PubMed]

35. Ruiz-Vera, U.M.; De Souza, A.P.; Long, S.P.; Ort, D.R. The role of sink strength and nitrogen availability in the down-regulation of photosynthetic capacity in field-grown *Nicotiana tabacum* L. at elevated CO_2 concentration. *Front. Plant Sci.* **2017**, *8*, 998. [CrossRef] [PubMed]

36. Oxborough, K.; Baker, N.R. Resolving chlorophyll a fluorescence images of photosynthetic efficiency into photochemical and non-photochemical components – calculation of qP and Fv'/Fm' without measuring Fo'. *Photosynth. Res.* **1997**, *54*, 135–142. [CrossRef]

37. Niinemets, Ü.; Kull, O.; Tenhunen, J.D. Within-canopy variation in the rate of development of photosynthetic capacity is proportional to integrated quantum flux density in temperate deciduous trees. *Plant Cell Environ.* **2004**, *27*, 293–313. [CrossRef]

38. Flexas, J.; Bota, J.; Galmés, J.; Medrano, H.; Ribas-Carbó, M. Keeping a positive carbon balance under adverse conditions: Responses of photosynthesis and respiration to water stress. *Physiol. Plant.* **2006**, *127*, 343–352. [CrossRef]

39. Kitao, M.; Lei, T.T. Circumvention of over-excitation of PSII by maintaining electron transport rate in leaves of four cotton genotypes developed under long-term drought. *Plant Biol.* **2007**, *9*, 69–76. [CrossRef]

40. Kitao, M.; Lei, T.T.; Koike, T.; Tobita, H.; Maruyama, Y. Higher electron transport rate observed at low intercellular CO_2 concentration in long-term drought-acclimated leaves of Japanese mountain birch (*Betula ermanii*). *Physiol. Plant.* **2003**, *118*, 406–413. [CrossRef]

41. Farquhar, G.D.; von Caemmerer, S. Modelling of Photosynthetic Response to Environmental Conditions. In *Physiological Plant Ecology II*; Springer: Berlin/Heidelberg, Germany, 1982; pp. 549–587.

42. Nagashima, T.; Sudo, K.; Akimoto, H.; Kurokawa, J.; Ohara, T. Long-term change in the source contribution to surface ozone over Japan. *Atmos. Chem. Phys.* **2017**, *17*, 8231–8246. [CrossRef]

43. Sicard, P.; Anav, A.; De Marco, A.; Paoletti, E. Projected global ground-level ozone impacts on vegetation under different emission and climate scenarios. *Atmos. Chem. Phys.* **2017**, *17*, 12177–12196. [CrossRef]

44. Karnosky, D.F.; Pregitzer, K.S.; Zak, D.R.; Kubiske, M.E.; Hendrey, G.R.; Weinstein, D.; Nosal, M.; Percy, K.E. Scaling ozone responses of forest trees to the ecosystem level in a changing climate. *Plant Cell Environ.* **2005**, *28*, 965–981. [CrossRef]

45. Ainsworth, E.A.; Long, S.P. What have we learned from 15 years of free-air CO_2 enrichment (FACE)? A meta-analytic review of the responses of photosynthesis, canopy properties and plant production to rising CO_2. *New Phytol.* **2005**, *165*, 351–372. [CrossRef]

46. Norby, R.J.; Zak, D.R. Ecological Lessons from Free-Air CO_2 Enrichment (FACE) Experiments. *Annu. Rev. Ecol. Evol. Syst.* **2011**, *42*, 181–203. [CrossRef]

47. Agathokleous, E.; Saitanis, C.J.; Wang, X.; Watanabe, M.; Koike, T. A review study on past 40 years of research on effects of tropospheric O_3 on belowground structure, functioning and processes of trees: A linkage with potential ecological implications. *Water Air Soil Pollut.* **2016**, *227*, 33. [CrossRef]

48. Watanabe, M.; Hoshika, Y.; Inada, N.; Wang, X.; Mao, Q.; Koike, T. Photosynthetic traits of Siebold's beech and oak saplings grown under free air ozone exposure in northern Japan. *Environ. Pollut.* **2013**, *174*, 50–56. [CrossRef] [PubMed]

49. Pell, E.J.; Schlagnhaufer, C.D.; Arteca, R.N. Ozone-induced oxidative stress: Mechanisms of action and reaction. *Physiol. Plant.* **1997**, *100*, 264–273. [CrossRef]

50. Agathokleous, E.; Kitao, M.; Calabrese, E.J. Environmental hormesis and its fundamental biological basis: Rewriting the history of toxicology. *Environ. Res.* **2018**, *165*, 274–278. [CrossRef]

51. Agathokleous, E.; Kitao, M.; Calabrese, E.J. Hormesis: A compelling platform for sophisticated plant science. *Trends Plant Sci.* **2019**, in press. [CrossRef]

52. Calabrese, E.J.; Agathokleous, E. Building biological shields via hormesis. *Trends Pharmacol. Sci.* **2019**, *40*, 8–10. [CrossRef]

53. Long, S.P.; Humphries, S.; Falkowski, P.G. Photoinhibition of Photosynthesis in Nature. *Annu. Rev. Plant Physiol. Plant Mol. Biol.* **1994**, *45*, 633–662. [CrossRef]

Article

Effects of Combined CO$_2$ and O$_3$ Exposures on Net CO$_2$ Assimilation and Biomass Allocation in Seedlings of the Late-Successional *Fagus Crenata*

Hiroyuki Tobita [1,*], Masabumi Komatsu [1], Hisanori Harayama [2], Kenichi Yazaki [1], Satoshi Kitaoka [1] and Mitsutoshi Kitao [2]

[1] Department of Plant Ecology, Forestry and Forest Products Research Institute, Matsunosato 1, Tsukuba 305-8687, Japan; kopine@ffpri.affrc.go.jp (M.K.); kyazaki@ffpri.affrc.go.jp (K.Y.); skitaoka3104@gmail.com (S.K.)

[2] Hokkaido Research Center, Forestry and Forest Products Research Institute, Hitsujigaoka 7, Sapporo 062-8516, Japan; harahisa@ffpri.affrc.go.jp (H.H.); kitao@ffpri.affrc.go.jp (M.K.)

* Correspondence: tobi@ffpri.affrc.go.jp

Received: 15 July 2019; Accepted: 21 September 2019; Published: 26 September 2019

Abstract: We examined the effects of elevated CO$_2$ and elevated O$_3$ concentrations on net CO$_2$ assimilation and growth of *Fagus crenata* in a screen-aided free-air concentration-enrichment (FACE) system. Seedlings were exposed to ambient air (control), elevated CO$_2$ (550 µmol mol^{-1} CO$_2$, +CO$_2$), elevated O$_3$ (double the control, +O$_3$), and the combination of elevated CO$_2$ and O$_3$ (+CO$_2$+O$_3$) for two growing seasons. The responses in light-saturated net CO$_2$ assimilation rates per leaf area ($A_{\text{growth-CO}_2}$) at each ambient CO$_2$ concentration to the elevated CO$_2$ and/or O$_3$ treatments varied widely with leaf age. In older leaves, $A_{\text{growth-CO}_2}$ was lower in the presence of +O$_3$ than in untreated controls, but +CO$_2$+O$_3$ treatment had no effect on $A_{\text{growth-CO}_2}$ compared with the +CO$_2$ treatment. Total plant biomass increased under conditions of elevated CO$_2$ and was largest in the +CO$_2$+O$_3$ treatment. Biomass allocation to roots decreased with elevated CO$_2$ and with elevated O$_3$. Elongation of second-flush shoots also increased in the presence of elevated CO$_2$ and was largest in the +CO$_2$+O$_3$ treatment. Collectively, these results suggest that conditions of elevated CO$_2$ and O$_3$ contribute to enhanced plant growth; reflecting changes in biomass allocation and mitigation of the negative impacts of O$_3$ on net CO$_2$ assimilation.

Keywords: allometric relationship; determinant species; leaf aging; stomatal conductance

---

1. Introduction

Atmospheric carbon dioxide (CO$_2$) concentrations are predicted to double during the next century, and tropospheric ozone (O$_3$) levels have also continued to rise globally since pre-industrial times [1–5]. Both CO$_2$ and O$_3$ are recognized as anthropogenic air pollutants with opposing impacts on plant growth [6–9]. Elevated CO$_2$ stimulates plant growth by enhancing photosynthetic carbon assimilation [10–12], although long-term exposure to elevated CO$_2$ results in photosynthetic down-regulation, especially under limiting nutrient conditions [13,14]. Elevated O$_3$ concentrations reduce net CO$_2$ assimilation (A) and accelerate leaf senescence [15,16], potentially resulting in substantial losses of the C-sink strengths in trees [9,17–20].

Elevated CO$_2$ may alleviate the negative effects of O$_3$ on photosynthetic activity and plant growth [8,21]. Free-air concentration enrichment (FACE) experiments for CO$_2$ and other ambient air experiments indicate a nonlinear interaction between plant responses to CO$_2$ and O$_3$ [7,22,23]. Hence, O$_3$ exposure may alter the C metabolism and decrease C stocks for tree species, likely through changes in quantities or the compositions of nonstructural carbohydrates [24]. They may in turn alter C allocation in plants [19,25]. Increased O_3 concentrations have been shown to shift carbon allocation in favor of aboveground plant tissues, especially leaves, at the expense of root growth, presumably to allocate resources for detoxification and repair of damaged leaves [6,16,26–30]. As elevated CO_2 can alleviate the O_3-induced reduction in *A*, shifts in biomass allocation into leaves may have an over-compensating effect on plant growth under the combination of elevated CO_2 and O_3. Although mechanisms of O_3-induced responses in plants have been revealed [31], responses to elevated O_3 vary widely among species [3,32]. Moreover, interactions between environmental variables such as elevated CO_2, drought, and high temperature, and their effects on crop productivity, are poorly understood [31].

Scaling up from the laboratory to the field with FACE technologies will be crucial to the understanding of plant responses to elevated O_3 and CO_2 [31]. Yet free-air exposure experiments with combinations of elevated CO_2 and O_3 [33] and with elevated O_3 only [17,34,35] are fewer than those performed in free air CO_2 experiments [36–39], especially in Asia [40]. In Japan, free-air CO_2 and O_3 exposure experiments have been conducted under field conditions [41–44]. Kitao et al., [45] reported the growth responses to elevated CO_2 and O_3 in two *Quercus* species, *Q. mongolica* and *Q. serrata*, which have relatively high tolerance to O_3 among Japanese tree species [46,47]. A significant enhancement of growth was observed in the seedlings grown under the combination of elevated CO_2 and O_3, compared with those grown under elevated CO_2 alone. This growth enhancement was considered as an over-compensating response, consisting of two processes: (1) preferential biomass allocation into leaves induced by elevated O_3, and (2) enhanced *A* by elevated CO_2.

Fagus crenata is a representative deciduous broadleaf tree species, widely distributing in the cool temperate forests in Japan [48]. This species is more susceptible to O_3 than the *Quercus* species described above [45]. Even at the current ambient O_3 levels, higher O_3 uptake by a forest of *F. crenata* resulted in an accelerated autumn senescence based on the observations by a flux tower [49]. Previous studies of plant-level *F. crenata* responses to interactions of CO_2 and O_3 produced contradictory results, with one study showing increased growth by combined elevated CO_2 and O_3 in greenhouse experiments [50], and another showing that elevated CO_2 concentrations did not mitigate the negative effects of elevated O_3 on plant growth [51]. Whether elevated CO_2 concentrations ameliorate the negative impacts of O_3 on growth is not certain, even in the same species [52]. To understand growth responses of *F. crenata* to elevated CO_2 and O_3, changes in C allocation into leaves and effects on *A* should be investigated collectively.

Quercus mongolica and *Q. serrata*, mid-successional species with a flush and succeeding-type of leaf emergence [53], can flush new shoots several times under ideal conditions. Unlike these *Quercus* species, *F. crenata*, a shade-tolerant late-successional species [48], has a flush type shoot developmental pattern [53], in which shoots are usually flushed only once in spring. We hypothesized that growth enhancement might be observed in the seedlings of *F. crenata* grown under the combination of elevated O_3 and CO_2, whereas the related growth enhancement is unlikely to be as large as those in the two *Quercus* species [45], because the leaf expansion pattern of *F. crenata* has lower plasticity to changes in environmental conditions. To test this hypothesis, we performed screen-aided FACE experiments and investigated the combined effects of elevated CO_2 and O_3 concentrations on *A* (as a leaf level trait) and biomass allocation (as a plant-level trait) in *F. crenata* seedlings without limiting root growth.

2. Material and Methods

2.1. Experimental Design

Screen-aided free-air concentration enrichment (FACE) experiments with elevated O_3 and CO_2 were performed in the facility used in previous studies [41,45]. The experimental field is located in the nursery of the Forestry and Forest Products Research Institute in Tsukuba, Japan (36°00′ N, 140°08′ E, 20 m a.s.l.). Plants were exposed to unchanged ambient air (control), elevated CO_2 (+CO_2; 550 μmol mol^{-1} CO_2), elevated O_3 (+O_3; twice-ambient), or elevated CO_2 and elevated O_3 (+CO_2+O_3; 550 μmol mol^{-1} CO_2 and twice-ambient O_3). A total of twelve frames (2 m width × 2.5 m height) were installed with three replicates for each treatment. CO_2 and O_3 were supplied during the daytime with a proportional integral derivative (PID) control system comprised of digital controllers (Model SDC35, Azbil Corporation, Tokyo, Japan). Gaseous CO_2 was obtained from liquid CO_2 (AIR WATER INC., Osaka, Japan). O_3 was supplied by an ozone generator (Model PZ2A; Kofloc, Kyoto, Japan). CO_2 concentration was measured by two CO_2 monitors (Carbon Dioxide Probe, Model GMP343; Vaisala, Helsinki, Finland); in addition, CO_2 concentration in the frames was monitored by an infrared CO_2 analyzer (Model LI-820; LI-COR Inc., Lincoln, NE, USA). Ozone concentration was measured by both an O_3 analyzer (Model EG-3000F; Ebara Jitsugyo Co. Ltd., Kanagawa, Japan) and an O_3 monitor (Model 205; 2B Technologies, Boulder, CO, USA). The exposure periods for CO_2 and O_3 extended from 20 May to 30 November in 2011, and from 25 April to 20 November in 2012 (these data are the same as those reported by Hiraoka et al. [41]). Mean ($\pm$ SE) daytime (7:00–17:00) CO_2 and O_3 concentrations during treatments were 36.1 ± 1.0 and 65.3 ± 2.0 ppb in 2011 and 36.4 ± 0.5 and 61.5 ± 6.2 ppb in 2012 for ambient and elevated O_3, respectively, and were 378 ± 2.5 and 562 ± 16.9 ppm in 2011 and 377 ± 2.9 and 546 ± 21.3 ppm in 2012 for ambient and elevated CO_2, respectively. Frames were surrounded by two transparent windscreens (50 cm height), encircled at 15 and 75 cm above the soil. The windscreens facilitated turbulent mixing of CO_2 and O_3, which were injected with ambient air through holes in vertical polyethylene tubes (2 m in height at 20-cm intervals around the frames) [54]. During the growth seasons from the beginning of April to the end of October in 2011 and 2012, mean monthly temperatures ranged from 8.2 °C to 28.4 °C. Precipitation for the growth period was 1096 mm in 2011 and 1019 mm in 2012 [41].

2.2. Plant Materials

Fagus crenata is a representative deciduous broadleaf tree species in Japan and is mainly distributed in deciduous broadleaf forests of the warm-temperate zone of East Asia. *Fagus crenata* have late-successional traits. Their shoot developmental pattern is classified as flush type shoot growth [53], in which shoots are usually flushed once in the spring. Nine current-year, dormant seedlings of *F. crenata* (about 5 cm in height) were transplanted directly into the ground in each frame in March 2011. Tree seedlings were grown for two growth seasons under each of the treatment conditions described above. The maximum height of the seedlings at the end of this experiment was 132.7 cm, which was lower than the frame height.

2.3. Gas-Exchange Measurements

Light-saturated net CO_2 assimilation rates per leaf area ($A_{\text{growth-CO}_2}$) in each ambient CO_2 concentration level (at 380 µmol mol^{-1} and 550 µmol mol^{-1} for ambient and elevated CO_2 conditions, respectively) were measured five times in first-flush leaves during mid-May, early July, early August, mid-September, and late October of the second year of CO_2 and O_3 exposures. Leaf photosynthetic parameters at ambient (380 µmol mol^{-1}) and elevated CO_2 (550 µmol mol^{-1}) concentrations were measured in three individual seedlings per treatment combination using a portable photosynthesis system (Li-6400, Li-Cor) and a leaf chamber fluorometer (Li-6400-40, Li-Cor). Measurements were taken with the block temperature set at 27 °C and at a saturating light intensity of 1500 µmol m^{-2} s^{-1}. Light was provided by a red/blue LED array with blue light comprising 10% of the total PFD. In preliminary studies, this light intensity was sufficient to saturate A in *F. crenata*. Gas-exchange measurements were conducted from 9:00 to 15:00. Prior to these measurements, sunscreens were placed over the frames to prevent direct sunlight from illuminating the leaves.

In August 2012, A-C_i relationships of the first flush leaves were determined and used to calculate maximum rates of carboxylation at 25 °C (V_{cmax25}). The A-C_i curves were determined in six steps with saturating light at 1500 µmol m^{-2} s^{-1}, starting from an ambient CO_2 concentration for each treatment (380 or 550) followed by 100-, 200-, 380-, 550-, and 1000-µmol CO_2 mol^{-1}. Measurements were taken at a block temperature of 27 °C. Maximum rates of carboxylation (V_{cmax}) were estimated using the following equation of the Farquhar-type model [55]: $A = V_{\text{cmax}} (C_c - \Gamma^*)/\{C_c + K_c \times (1 + O_i/K_o)\}$

$- R_d$, where C_c is the CO_2 concentration in the chloroplasts, Γ^* is the CO_2 compensation point, K_c and K_o are the Michaelis–Menten constants for carboxylation and oxygenation, O_i is the intercellular O_2 concentration, and R_d is the rate of daytime respiration. In this study, we ignored possible CO_2 diffusion limitations within leaves, and used C_i as a proxy for C_c. Values for the coefficients K_c, K_o, and Γ^* were 404.9, 278.4, and 42.75 μmol mol^{-1}, respectively, at 25 °C [56]. V_{cmax} was calculated by fitting the equation to the initial slope of the A/C_i curve ($C_i < 300$).

2.4. Chlorophyll Fluorescence Measurements

To quantitate the degree of aging-related senescence following CO_2 and O_3 treatments, we determined initial (F_0) and maximum fluorescence (F_m) values in all leaves used for the gas exchange measurements after overnight dark-adaptation using a portable fluorometer (MINI-PAM, Walz, Effeltrich, Germany). Saturating pulse of 7000 mol m^{-2} s^{-1} for 1 s was used to determine F_m values. The ratio of variable to maximum fluorescence (F_v/F_m where $F_v = F_m - F_0$) represents the maximum PS II photochemical efficiency [57].

2.5. Leaf Characteristics

Leaves were collected after the measurements of F_v/F_m and leaf areas were measured using a scanner and Image J software (NIH, Bethesda, MD, USA). Dry weights of each leaf were measured after oven-drying at 70 °C and leaf mass per leaf area (LMA) was calculated for each leaf. Nitrogen concentrations of each leaf (leaf N_{mass}) were analyzed using an NC-analyzer (NT-900, SUMIKA Chem., Osaka, Japan), and N contents per leaf area (leaf N_{area}) were calculated using LMA.

2.6. Growth Measurements

In November of the second year, gas-exchange measurements were completed and all seedlings were harvested ($n = 15$, control; $n = 15$, +CO_2; $n = 22$, +O_3; $n = 21$, +CO_2+O_3). Root systems were then excavated using an air excavation tool (Air Schop, KF Company, Sakai, Japan) to loosen soils [45]. Loosened soils were then carefully removed by hand using small stainless-steel rakes (Bonsai rake, Kikuwa, Sanjo, Niigata, Japan). Dry weights of leaves, current-year shoots, stems, and roots were determined after oven-drying to a constant weight at 70 °C.

2.7. Statistical Analysis

Significant effects of CO_2 and O_3 treatments on plant biomass, organ biomass (leaves, current-year shoots, stems and non-current-year shoots, and roots), second-flush shoot lengths, $A_{growth\text{-}CO_2}$, $g_{sw\text{-}growth\text{-}CO_2}$, C_i, leaf N_{mass}, leaf N_{area}, and LMA were identified using ANOVA with mean values from each frame ($CO_2 \times O_3$ regimen; $n = 12$). Values were averaged to calculate sample estimates for each replicate. Standardized major axis tests and routines (SMA) regression [58,59] analyses were used to examine relationships among log$_{10}$-transformed whole plant and plant tissue (g plant^{-1}) biomasses using the SMATR 3.3 package [60]. Differences in elevations of SMA regressions were tested among treatments using Sidak's adjusted pair-wise tests, depending on the significance of the null hypothesis that the slopes were equal. Analyses were performed using data from all plants of each treatment group. All statistical analyses were performed using R version 3.2.3 [61].

3. Results

3.1. Photosynthesis

Net CO_2 assimilation rate responses to CO_2 and O_3 treatments varied widely with leaf age (Figure 1A, Table 1). Although $A_{growth\text{-}CO_2}$ increased following treatment with elevated CO_2 concentrations in younger leaves (May, $p = 0.088$; July, $P = 0.082$), no positive effects of elevated CO_2 were observed in older leaves (August to October). During August, V_{cmax25} declined more in the +O_3 treatment group than in control plants, yet no significant effects of CO_2 and O_3 treatments were identified (Table 1). In October, $A_{growth\text{-}CO_2}$ did not significantly decline due to elevated O_3 exposure, and predawn F_v/F_m was lower in the $+O_3$ treatment group than in all other groups (Table 1). Stomatal conductance ($g_{sw\text{-}growth\text{-}CO_2}$) decreased by elevated CO_2 in May ($p = 0.036$) and July ($p = 0.017$) (Figure 1B, Table 1). A tendency for effects of the O_3 were also detected only in July ($p = 0.058$). During September and October, seedlings of the $+CO_2+O_3$ group tended to have the lowest $g_{sw\text{-}growth\text{-}CO_2}$ values. Finally, elevated O_3 had no effect on C_i throughout the growing season (Figure 1C).

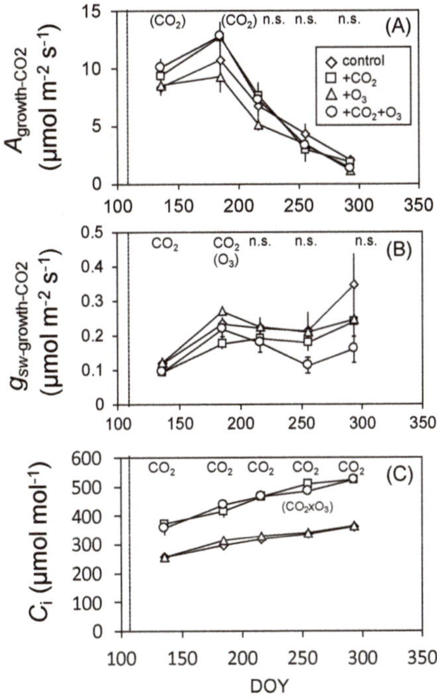

Figure 1. Seasonal changes in the light-saturated net CO_2 assimilation rate ($A_{growth\text{-}CO_2}$) (**A**), stomatal conductance ($g_{sw\text{-}growth\text{-}CO_2}$) (**B**), and intercellular CO_2 concentration (C_i) (**C**) in the first flush leaves of *F. crenata* seedlings grown under the CO_2 and O_3 treatment combinations at each ambient CO_2 level (380 and 550 μmol mol^{-1}). Diamond: control, square: elevated CO_2, triangle: elevated O_3, circle: the combination of elevated CO_2 and O_3. Dotted line shows the period of the bud break. The effects ($p < 0.05$) of CO_2 treatments, O_3 treatments, and the interaction CO_2 and O_3 on each parameter for each measurement period (five times) are indicated in the panel by CO_2, O_3, and $CO_2 \times O_3$, respectively. The parenthesis indicate difference at $p < 0.1$, and n.s. indicate nonsignificant ($p > 0.1$). Values are means $\pm$ SE ($n = 3$).

3.2. Leaf Characteristics

Leaf N_{mass} of *F. crenata* seedlings was decreased by CO_2 treatment in May ($F_{1,8} = 25.8$; $p = 0.001$), and August ($F_{1,8} = 3.8$; $p = 0.088$; Figure 2A). In August, O_3 treatments also affected leaf N_{mass} ($F_{1,8} = 9.2$; $p = 0.016$). Effects of CO_2 in May ($F_{1,8} = 6.9$; $p = 0.031$) and its interactions with O_3 in July ($F_{1,8} = 4.0$; $p = 0.080$) on LMA were observed (Figure 2B). Specifically, $+CO_2+O_3$-treated seedlings had the highest LMA values among the four treatment groups in all growth periods, yet no significant effects of CO_2 and O_3 treatments were identified. Leaf N_{area} values showed interactive effects of CO_2 and O_3 treatments in September ($F_{1,8} = 7.1$; $p = 0.028$), May ($F_{1,8} = 3.8$; $p = 0.087$) and August ($F_{1,8} = 3.7$; $p = 0.091$) in addition to the effects of O_3 treatments in August (Figure 2C; $F_{1,8} = 21.5$; $p = 0.0017$).

Table 1. Light-saturated net CO_2 assimilation rate ($A_{growth-CO_2}$) and stomatal conductance ($g_{sw-growth-CO_2}$) in May and July at each ambient CO_2 level (380 and 550 μmol mol^{-1}), the maximum rate of carboxylation at 25 °C (V_{cmax25}) in August, and the maximum PS II photochemical efficiency (F_v/F_m) at predawn in October in the first flush leaves of *F. crenata* grown under ambient air (control), elevated CO_2 (+CO_2), elevated O_3 (+O_3), and the combination of elevated CO_2 and O_3 (+CO_2+O_3). Values at upper half lines are means ± SE of three replicates for each treatment. *F* values of analysis of variance (ANOVA) to test the main effects of CO_2, O_3, and their interactions are also shown at lower half lines. Significant effects are indicated by *; $p < 0.05$, and n.s.; non-significant.

Treatments	$A_{growth-CO_2}$ (μmol m^{-2} s^{-1})		$g_{sw-growth-CO_2}$ (μmol m^{-2} s^{-1})		V_{cmax25} (μmol m^{-2} s^{-1})	Predawn-F_v/F_m
	(May)	(July)	(May)	(July)	(August)	(October)
Control	8.4 ± 0.8	10.8 ± 1.6	0.12 ± 0.01	0.23 ± 0.02	30.1 ± 3.5	0.76 ± 0.01
+CO_2	9.4 ± 0.3	12.7 ± 1.3	0.09 ± 0.01	0.18 ± 0.02	23.0 ± 2.1	0.78 ± 0.01
+O_3	8.6 ± 0.6	9.3 ± 1.4	0.12 ± 0.01	0.27 ± 0.01	21.9 ± 2.0	0.65 ± 0.04
+CO_2+O_3	10.1 ± 0.7	12.9 ± 1.1	0.09 ± 0.01	0.22 ± 0.02	22.2 ± 2.8	0.72± 0.01
Effect						
CO_2 ($F_{1,8}$)	3.7 n.s.	4.0 n.s.	6.34 *	8.94 *	1.65 n.s.	0.08 n.s.
O_3 ($F_{1,8}$)	0.5 n.s.	0.2 n.s.	0.05 n.s.	4.90 n.s.	2.86 n.s.	2.97 n.s.
$CO_2 \times O_3$ ($F_{1,8}$)	0.2 n.s.	0.3 n.s.	0.00 n.s.	0.01 n.s.	1.93 n.s.	0.03 n.s.

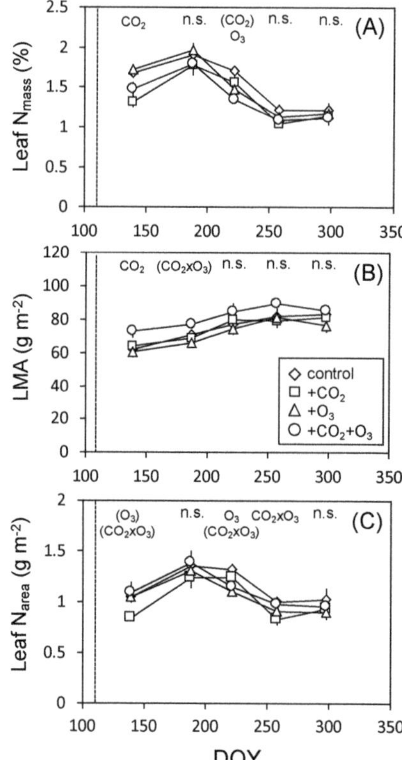

Figure 2. Seasonal change in the leaf N concentration (leaf N$_{mass}$) (**A**), leaf mass per area (LMA) (**B**), and area-based leaf N content (leaf N$_{mass}$) (**C**) in the first flush leaves of *F. crenata* seedlings grown under the CO_2 and O_3 treatment combinations at each ambient CO_2 level (380 and 550 μmol mol^{-1}). The format of these figures is the same as in Figure 1.

3.3. Plant Biomass

Elevated CO_2 concentrations had positive effects on the whole plant biomasses (Total) of *F. crenata* seedlings (Table 2; $p = 0.033$). Seedlings of the $+CO_2+O_3$ treatment group tended to exhibit the largest average plant biomass among the four treatment groups ($p = 0.037$, in one-way ANOVA). Though no significant effects of CO_2 and O_3 treatments were identified in dry matter distribution, elevated CO_2 concentrations showed a tendency to decrease RWR ($p = 0.078$) and increase aboveground biomass to root ratio (S:R ratio) ($p = 0.066$), and leaf to root ratio (L:R ratio) tended to increase by elevated O_3 treatment ($p = 0.065$) (Table 3). All treatment groups included some seedlings that had a second flush leaf production. Yet only following $+CO_2+O_3$ treatments did all plants produce second-flush leaves (77.8% ± 14.7%, Control; 68.9% ± 17.4%, $+CO_2$; 81.9% ± 10.8%, $+O_3$; and 100%, $+CO_2+O_3$; these values indicated mean ± SE). Lengths of second-flush shoots (mean ± SE) were also affected by CO_2 concentrations ($p = 0.030$) and were 31.4 ± 12.4 cm in control plants, 51.3 ± 5.9 cm in $+CO_2$ treated plants; 32.7 ± 2.9 cm in $+O_3$ treated plants, and 49.8 ± 1.1 cm in $+CO_2+O_3$ treated plants.

Table 2. Dry mass of plant tissues (whole plant; Total, leaf weight; Leaf, current shoot weight; Current shoot; stem and shoot weight; Stem and shoot, root weight; Root) of the seedling of *F. crenata* grown under ambient air (control), elevated CO_2 ($+CO_2$), elevated O_3 ($+O_3$), and the combination of elevated CO_2 and O_3 ($+CO_2+O_3$). Values at upper half lines are means ± SE of three replicates for each treatment. *F* values of analysis of variance (ANOVA) to test the main effects of CO_2, O_3, and their interactions are also shown at lower half lines. Significant effects are indicated by *; $p < 0.05$, and n.s.; non-significant.

Treatments	Total	Leaf	Current Shoot	Stem and Shoot	Root
Control	43.3 ± 9.6	5.5 ± 1.5	8.8 ± 3.2	8.7 ± 1.3	20.3 ± 3.7
$+CO_2$	45.4 ± 6.4	5.3 ± 1.3	10.0 ± 2.6	10.2 ± 1.6	19.9 ± 2.2
$+O_3$	39.2 ± 0.7	5.4 ± 0.5	6.9 ± 0.3	8.9 ± 1.2	18.0 ± 0.9
$+CO_2+O_3$	67.1 ± 1.3	9.2 ± 0.4	16.4 ± 0.6	15.1 ± 1.3	26.4 ± 0.8
Effect					
CO_2 ($F_{1,8}$)	6.6 *	3.1 n.s.	6.5 *	7.8 *	3.2 n.s.
O_3 ($F_{1,8}$)	2.3 n.s.	3.4 n.s.	1.2 n.s.	3.5 n.s.	0.9 n.s.
$CO_2 \times O_3$ ($F_{1,8}$)	4.9 n.s.	3.8 n.s.	3.9 n.s.	2.9 n.s.	3.8 n.s.

Table 3. Dry matter distribution (leaf weight ratio; LWR, current shoot weight ratio; CSWR, stem and shoot weight ratio; SWR, root weight ratio; RWR, and aboveground biomass to root ratio; S:R ratio, leaf to root ratio; L:R ratio) of the seedling of *F. crenata* grown under ambient air (control), elevated CO_2 ($+CO_2$), elevated O_3 ($+O_3$), and the combination of elevated CO_2 and O_3 ($+CO_2+O_3$). Values at upper half lines are means ± SE of three replicates for each treatment. *F* values of analysis of variance (ANOVA) to test the main effects of CO_2, O_3, and their interactions are also shown at lower half lines. There is no significant effect in all parameters, which indicated by n.s.; non-significant.

Treatments	LWR	CSWR	SWR	RWR	S:R Ratio	L:R Ratio
Control	0.12 ± 0.01	0.19 ± 0.03	0.21 ± 0.02	0.48 ± 0.02	1.12 ± 0.09	0.26 ± 0.04
$+CO_2$	0.12 ± 0.02	0.21 ± 0.03	0.23 ± 0.03	0.44 ± 0.02	1.30 ± 0.11	0.26 ± 0.04
$+O_3$	0.14 ± 0.01	0.18 ± 0.00	0.23 ± 0.03	0.45 ± 0.02	1.25 ± 0.13	0.30 ± 0.02
$+CO_2+O_3$	0.14 ± 0.01	0.24 ± 0.01	0.21 ± 0.01	0.41 ± 0.01	1.50 ± 0.06	0.35 ± 0.02
Effect						
CO_2 ($F_{1,8}$)	0.01 n.s.	2.84 n.s.	0.03 n.s.	4.07 n.s.	4.52 n.s.	0.82 n.s.
O_3 ($F_{1,8}$)	3.37 n.s.	0.09 n.s.	0.01 n.s.	2.24 n.s.	2.65 n.s.	4.57 n.s.
$CO_2 \times O_3$ ($F_{1,8}$)	0.05 n.s.	0.82 n.s.	0.65 n.s.	0.08 n.s.	0.18 n.s.	0.91 n.s.

3.4. Biomass Allocation

In our allometric assessments of relationships between whole plant weights and root weights, regression slopes did not differ among treatments ($p = 0.231$; Figure 3A, Table 4). In case the common slopes of 1.015 (0.947–1.091; ±95% confidence intervals (CI)) were used, magnitudes of relationships differed significantly between treatment groups (p = 0.0002). Specifically, seedlings from the $+CO_2+O_3$ treatment group had smaller elevations than those of the control and the $+O_3$ treatment groups. Allometric assessments of root and leaf weights also gave slopes that did not differ between treatments (p = 0.114; Figure 3B, Table 4). At common slopes of 0.979 (0.831–1.153; ± 95% CI), magnitudes of these relationships differed significantly between treatment groups (p = 0.0087). The $+CO_2+O_3$ treatment led to greater elevation than that observed in control and the $+CO_2$ treatment group. In allometric assessments, the regression slopes in relationships between whole plant and leaf weights (p = 0.013) and between whole plant and shoot weights (p = 0.032) differed significantly between treatments. The slopes (p = 0.635) and magnitudes (p = 0.725) of the allometric relationships between whole plant and stem weights did not differ.

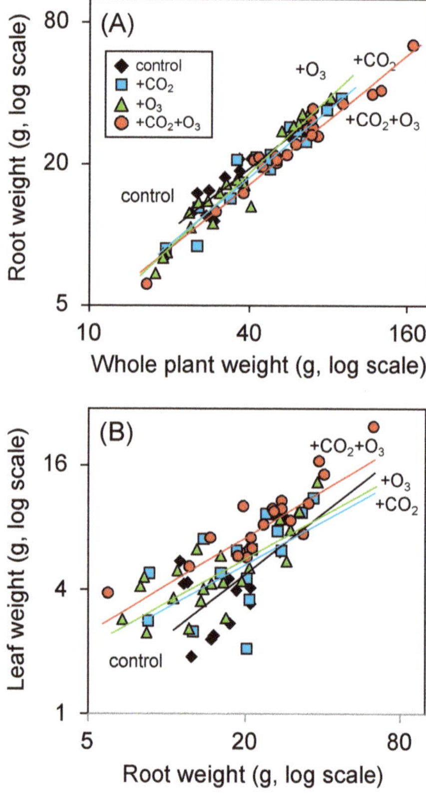

Figure 3. Allometric relationships between whole plant weight and root weight (**A**) and between root weight and leaf weight (**B**) of *F. crenata* seedlings. Allometric equations analyzed by using an R package, smart 3 (SMA analysis), are listed in Table 4. Diamond (black): control, square (blue): elevated CO_2, triangle (green): elevated O_3, circle (red): the combination of elevated CO_2 and O_3. Each colored line shows regression line for each treatment.

Table 4. Standardized major axis (SMA) regression slopes and their confidence intervals for log-log transformed relationships between whole plant weight and root weight (Figure 3A) and between root weight and leaf weight (Figure 3B) of *F. crenata* seedlings grown under ambient air (control), elevated CO_2 (+CO_2), elevated O_3 (+O_3), and the combination of elevated CO_2 and O_3 (+CO_2+O_3). Values of coefficients of determination (r^2) and significant values (p) of each bivariate relationship are shown. 95% confidence intervals (CI) of SMA slopes and y-axis intercepts are exhibited. In cases where SMA tests for common slopes revealed no significant differences between treatments, common slopes are used. The differences in elevations of the SMA regressions are tested among treatments using Sidak's adjusted pair-wise test, depending on the significance of the null hypothesis that the slopes are equal. Where there is a significant difference in elevation of the common-slope SMA regressions, values for the y-axis intercept (elevation) are provided. Different letters in pairwise comparison are significantly different, with a $p < 0.05$ under the slopes are equal. Where appropriate, significant shifts along a common slope are indicated. The format of table is referred that in Atkin et al. [59].

Response	Whole Plant Weight				Leaf Weight			
Bivariate	Root Weight				Root Weight			
H0 NO. 1: no difference in slope (p-value)		0.231					0.114	
Treatment	Control	+CO_2	+O_3	+CO_2+O_3	Control	+CO_2	+O_3	+CO_2+O_3
n	15	15	22	21	15	15	22	21
r^2	0.859	0.917	0.946	0.952	0.393	0.367	0.626	0.751
p-value	<0.0001	<0.0001	<0.0001	<0.0001	0.012	0.017	<0.0001	<0.0001
Slope	0.961	1.018	1.107	0.952	1.579	1.152	0.891	0.887
Slope CI_high	1.202	1.208	1.233	1.058	2.480	1.826	1.180	1.125
Slope CI_Low	0.768	0.858	0.993	0.857	1.006	0.727	0.672	0.700
Intercept	−0.257	−0.392	−0.509	−0.315	−0.592	−0.766	−0.385	−0.304
H0 NO. 2: no difference in elevation (p-value)		0.0002					0.009	
Intercepts for a common slope	−0.343	−0.387	−0.368	−0.427	−0.592	−0.544	−0.492	−0.431
Pairwise comparison (where relationship significant)								
	a	ab	a	b	b	b	ab	a
H0 NO. 3: no difference in 'shift' (p-value)		0.015					0.002	

4. Discussion

The present screen-aided FACE experiment demonstrated that the growth of *F. crenata* was enhanced by O_3 exposure under elevated CO_2, but not under ambient CO_2 (Table 2). The increased biomass of *F. crenata* seedling under +CO_2+O_3 treatment might reflect (1) an alleviation of the negative effect of O_3 on photosynthetic activity (leaf level response; Figure 1, Table 1), and (2) an increase in biomass allocation into leaves at the expense of root growth (plant-level responses; Figure 3, Table 4). The incremental increase in the biomass of *F. crenata* under +CO_2+O_3 was less than those reported in two *Quercus* species grown under elevated CO_2 and O_3 [45]. Watanabe et al. [50] also reported incremental increases in plant biomass of *F. crenata* in open top chamber (OTC) experiments with elevated CO_2 and O_3 concentrations. Our field experiment confirmed that elevated CO_2 ameliorated the negative effects of elevated O_3 in *F. crenata* trees, which are highly susceptible to increased O_3 levels.

The positive effects by elevated CO_2 on the whole plant biomasses of *F. crenata* seedlings were shown in this relatively short-term experiment (Table 2). It is obvious that the responses to elevated CO_2 are dependent on the duration of exposure. Agathokleous et al., [62] reported that 11 years of exposure of CO_2 in the FACE system, established in the nursery of Hokkaido University (Sapporo, Japan), induced rhizomorphogenesis, with a massive production of fine roots, especially in the phosphorus-limited volcanic ash soils, but not in the brown forest soils. The observations from different systems and experimental setups suggest that the effect of elevated CO_2 on *F. crenata* would be dependent on soil conditions.

Fagus crenata seedlings grown under $+CO_2+O_3$ showed the lowest $g_{sw\text{-growth-}CO_2}$ values among the all treatment combinations in the later growing season (Figure 1B), suggesting a reduction in O_3 uptake under the conditions of $+CO_2+O_3$. Hence, as a leaf level photosynthetic response, the negative effects of O_3 on $A_{growth\text{-}CO_2}$ of *F. crenata* might be alleviated by higher CO_2 levels due to reduced O_3 uptake (Figure 1A). Elevated CO_2–induced decreases in stomatal conductance were also reported in *F. crenata* and other species under high CO_2 and O_3 levels [50,51]. Compensation for O_3-induced reductions in area-based CO_2 uptake rates may partly reflect lowered O_3 uptake rates due to stomatal narrowing in response to elevated CO_2 [21,52]. Several studies on broadleaf tree species have reported similar effects of elevated CO_2 alleviating O_3-related foliar injury [21,51].

Plants grown under $+CO_2+O_3$ showed the highest LMA throughout growth season (Figure 2B). Plants with higher LMA reportedly have high tolerance to elevated O_3 [3,15,27], which may also decrease concentrations of total nonstructural carbohydrates (TNC; soluble sugar and starch) and alter their distributions among plant tissues [24]. On contrast, TNC is usually increased in leaves under elevated CO_2 conditions, which enhance LMA [63]. Ozone damages leaf cells and thus adversely affects plant production, reduces photosynthetic rates, and demands increased resource allocation to detoxify and repair leaves [6]. Lindroth [8] suggested that enriched in CO_2 concentrations can increase carbon availability for the production of defense compounds pathways that are upregulated in response to O_3. Matsumura et al. [51] and Karonen et al. [64] showed that condensed tannins are present at significantly higher levels in seedlings of *F. crenata* after combined treatments with high CO_2 and O_3 levels. The changes in TNC and these defense compounds may contribute to increased LMA in *F. crenata* under combined increases in CO_2 and O_3 levels, yet we did not determine both of them in the present study.

Elevated O_3 can impair C metabolism and decrease C stocks [24], and several studies indicate decreases in biomass allocation into roots [26]. In the plant-level responses of *F. crenata* seedlings shown herein, allometric relationships between whole plant weight and root weight, and between roots and leaves were modified by $+CO_2+O_3$ treatments (Figure 3, Table 4). This suggests that *F. crenata* seedlings grown under the $+CO_2+O_3$ allocated larger amounts of biomass to aboveground tissues than belowground tissues, resulting in higher S:R ratio and L:R ratio, and in lower RWR compared with other treatments (Table 3). Reduced C allocation into root has also been demonstrated in young [16,30,65] and mature trees of *Fagus* species under conditions of elevated O_3 [19]. However, in mature *Fagus* species, different responses to elevated O_3 were also observed, such as stimulation of root growth [66,67] and no changes in carbon allocation into belowground tissues [68]. In addition, soil nutrient supply mitigated the negative impacts of O_3 on biomass allocation in *F. crenata* seedlings [69,70].

In our study, all *F. crenata* seedlings grown under $+CO_2+O_3$ produced second-flush shoots. In addition, $+CO_2+O_3$ led to the greatest average length of the secondary shoots among the treatment combinations, although *F. crenata* is a determinant tree species that usually flushes shoots once a year. Watanabe et al. [50] also showed similar growth responses with enhanced second flushes of *F. crenata* under elevated CO_2 and O_3 in OTC experiments. This suggests that the greater investment of carbohydrate due to the higher photosynthetic rate in first flush leaves contributed to the increase in new leaf emergence. Kolb and Matyssek [71] similarly showed that plants with more determinate growth habits tend to respond to O_3 exposure by reducing carbon allocation to root growth, in favor of maintaining older leaves and supporting new leaf flushes. Sitch et al. [20] also suggested that translocation patterns to different plant organs might depend as much on sink activities as well as source strengths. In our study, the balance between higher source activity ($A_{growth\text{-}CO_2}$) due to increased CO_2 concentration, and higher sink strength for detoxification and repair of leaves following O_3 exposure, might be responsible for the changes in allometric relationships between roots and leaves. Conversely, lower $A_{growth\text{-}CO_2}$ and smaller numbers of second-flush leaves in *F. crenata* seedlings of the $+O_3$ treatment group may underlie the comparatively unclear changes in biomass allocation.

We observed declines in F_v/F_m ratios of *F. crenata* under elevated O_3 later in the growth season (Table 1), suggesting that autumn leaf senescence was accelerated by O_3 exposure. Watanabe et al. [72]

performed FACE experiments and showed ozone-induced reductions in light-saturated photosynthetic rates in upper canopy leaves of *F. crenata* during August and October, but not during July. Grams et al. [21] also reported marked declines in photosynthetic rate responses to O_3 in *Fagus sylvatica* during the late season. In addition, even current ambient atmospheric O_3 levels have accelerated autumn senescence in a forest of *F. crenata* in the northern part of Japan [49]. These dynamic interactions between plant development, carbon allocation, and O_3 exposure are important for understanding future impacts of ozone on forest ecosystems [6]. The results of our screen-aided free-air O_3 and CO_2 exposure experiments in the field may contribute to predictions of tree responses to future climate change.

Since both CO_2 and O_3 are categorized global-warming-gases, along with increased O_3 and CO_2 concentrations, ambient air temperature may increase in the future. However, our study did not treat air temperature, and the air temperature in each frame was not affected by CO_2 and O_3 treatments. Hence, this study focused on the effects of increased CO_2 and O_3 without consideration of temperature.

5. Conclusions

Screen-aided FACE experiments demonstrated that the growth of *Fagus crenata* seedlings was enhanced by combined increases in O_3 and CO_2 concentrations, which was related to the modified C allocation between roots and leaves (plant-level responses), and the alleviation of O_3 impact on net CO_2 assimilation by elevated CO_2 (leaf level responses). It is noteworthy that the extent of growth enhancement was not as large as that reported in *Quercus* species [45]. The intrinsic lower plasticity of leaf expansion pattern in *F. crenata* would affect the growth responses to future combinations of increased O_3 and CO_2 levels.

Author Contributions: H.T. and M.K. designed the study. H.T., M.K., H.H., K.Y., S.K., and M.K. collected the photosynthetic data, performed the analysis, and hence equally contributed to this study. All authors also discussed the results and commented on the manuscript.

Funding: This study was supported in part by the project on "Technology development for circulatory food production systems responsive to climate change" conducted by the Ministry of Agriculture, Forestry and Fisheries, Japan, the Grant-in-Aid for Scientific Research (B) (No. 25292092) and JSPS KAKENHI Grant Number JP17H03839.

Acknowledgments: We thank K Sakai and K. Arai for assistance with measurements. We also thank Morikawa for his valuable suggestions concerning the present study.

Conflicts of Interest: The authors declare no conflict of interest.

References

1. Chen, Z.; Shang, H.; Cao, J.; Yu, H. Effects of ambient ozone concentrations on contents of nonstructural carbohydrates in *Phoebe bournei* and *Pinus massoniana* seedlings in subtropical China. *Water Air Soil Pollut.* **2015**, *226*, 310–317. [CrossRef]

2. Dentener, F.; Stevenson, D.; Ellingsen, K.; van Noije, T.; Schultz, M.; Amann, M.; Atherton, C.; Bell, N.; Bergmann, D.; Bey, I.; et al. The global atmospheric environment for the next generation. *Environ. Sci. Technol.* **2006**, *40*, 3586–3594. [CrossRef] [PubMed]

3. Feng, Z.; Büker, P.; Pleijel, H.; Emberson, L.; Karlsson, P.E.; Uddling, J. A unifying explanation for variation in ozone sensitivity among woody plants. *Glob. Chang. Biol.* **2018**, *24*, 78–84. [CrossRef] [PubMed]

4. IPCC. *Climate Change 2013: The Physical Science Basis*; Stocker, T.F., Qin, D., Plattner, G.-K., Tignor, M., Allen, S.K., Boschung, J., Nauels, A., Xia, Y., Bex, B., Midgley, P.M., Eds.; Contribution of Working Group I to the Fifth Assessment Report of the Intergovernmental Panel on Climate Change; Cambridge University Press: Cambridge, UK, 2013.

5. Monks, P.S.; Archibald, A.T.; Colette, A.; Cooper, O.; Coyle, M.; Derwent, R.; Fowler, D.; Granier, C.; Law, K.S.; Mills, G.E.; et al. Tropospheric ozone and its precursors from the urban to the global scale from air quality to short-lived climate forcer. *Atmos. Chem. Phys.* **2015**, *15*, 8889–8973. [CrossRef]

6. Ashmore, M.R. Assessing the future global impacts of ozone on vegetation. *Plant Cell Environ.* **2005**, *28*, 949–964. [CrossRef]

7.   Karnosky, D.F.; Pregitzer, K.S.; Zak, D.R.; Kubiske, M.E.; Hendrey, G.R.; Weinstein, D.; Nosal, M.; Percy, K.E. Scaling ozone responses of forest trees to the ecosystem level in a changing climate. *Plant Cell Environ.* **2005**, *28*, 965–981. [CrossRef]

8.   Lindroth, R.L. Atmospheric change, plant secondary metabolites and ecological interactions. In *The Ecology of Plant Secondary Metabolites: From Genes to Global Progresses*; Iason, G.R., Dick, M., Hartley, S.E., Eds.; Cambridge University Press: Cambridge, UK, 2012; pp. 120–153.

9.   Wittig, V.E.; Ainsworth, E.A.; Naidu, S.L.; Karnosky, D.F.; Long, S.P. Quantifying the impact of current and future tropospheric ozone on tree biomass, growth, physiology and biochemistry: A quantitative meta-analysis. *Glob. Chang. Biol.* **2009**, *15*, 396–424. [CrossRef]

10. Ainsworth, E.A.; Davey, P.A.; Hymus, G.J.; Osborne, C.P.; Roger, A.; Blum, H.; Nösberger, J.; Long, S.P. Is stimulation of leaf photosynthesis by elevated carbon dioxide concentration maintained in the long term? A test with *Lolium perenne* grow for 10 years at two nitrogen fertilization levels under *Free Air* CO_2 Enrichment (FACE). *Plant Cell Environ.* **2003**, *26*, 705–714. [CrossRef]

11. Davey, P.A.; Olcer, H.; Zakhleniuk, O.; Bernacchi, C.J.; Calfapietra, C.; Long, S.P.; Raines, C.A. Can fast-growing plantation trees escape biochemical down-regulation of photosynthesis when grown throughout their complete production cycle in the open air under elevated carbon dioxide? *Plant Cell Environ.* **2006**, *29*, 1235–1244. [CrossRef]

12. Zak, D.R.; Pregitzer, K.S.; Kubiske, M.E.; Burton, A.J. Forest productivity under elevated CO_2 and O_3: Positive feedbacks to soil N cycling sustain decade-long net primary productivity enhancement by CO_2. *Ecol. Lett.* **2011**, *14*, 1220–1226. [CrossRef]

13. Norby, R.J.; Warren, J.M.; Iversen, C.M.; Medlyn, B.E.; McMurtrie, R.E. CO_2 enhancement of forest productivity constrained by limited nitrogen availability. *Proc. Natl. Acad. Sci. USA* **2010**, *107*, 19368–19373. [CrossRef] [PubMed]

14. Sigurdsson, B.D.; Medhurst, J.L.; Wallin, G.; Eggertsson, O.; Linder, S. Growth of mature boreal Norway spruce was not affected by elevated $[CO_2]$ and/or air temperature unless nutrient availability was improved. *Tree Physiol.* **2013**, *33*, 1192–1205. [CrossRef] [PubMed]

15. Li, P.; Calatayud, V.; Gao, F.; Uddling, J.; Feng, Z. Differences in ozone sensitivity among woody species are related to leaf morphology and antioxidant levels. *Tree Physiol.* **2016**, *36*, 1105–1116. [CrossRef] [PubMed]

16. Matyssek, R.; Sandermann, H. Impact of ozone on trees: An ecophysiological perspective. *Prog. Bot.* **2003**, *64*, 349–404.

17. Matyssek, R.; Wieser, G.; Ceulemans, R.; Rennenberg, H.; Pretzsch, H.; Haberer, K.; Low, M.; Nunn, A.J.; Werner, H.; Wipfler, P.; et al. Enhanced ozone strongly reduces carbon sink strength of adult beech (*Fagus sylvatica*)-Resume from the free-air fumigation study at Kranzberg forest. *Environ. Pollut.* **2010**, *158*, 2527–2532. [CrossRef] [PubMed]

18. Pretzsch, H.; Dieler, J.; Matyssek, R.; Wipfler, P. Tree and stand growth of mature Norway spruce and European beech under long-term ozone fumigation. *Environ. Pollut.* **2010**, *158*, 1061–1070. [CrossRef]

19. Ritter, W.; Andersen, C.P.; Matyssek, R.; Grams, T.E.E. Carbon flux to woody tissues in a beech/spruce forest during summer and in response to chronic O_3 exposure. *Biogeosciences* **2011**, *8*, 3127–3138. [CrossRef]

20. Sitch, S.; Cox, P.M.; Collins, W.J.; Huntingford, C. Indirect radiative forcing of climate change through ozone effects on the land-carbon sink. *Nature* **2007**, *448*, 791–795. [CrossRef] [PubMed]

21. Grams, T.E.E.; Anegg, S.; Haberle, K.; Langebartlet, C.; Matyssek, R. Interactions of chronic exposure to elevated CO_2 and O_3 level in the photosynthetic light and dark reactions of European beech (*Fagus sylvatica*). *New Phytol.* **1999**, *144*, 95–107. [CrossRef]

22. Percy, K.E.; Nosal, M.; Heilman, W.; Dann, T.; Sobre, J.; Legge, A.H.; Karnosky, D.F. New exposure-based metric approach for evaluating O_3 risk to North American aspen forests. *Environ. Pollut.* **2007**, *147*, 554–566. [CrossRef]

23. Volk, M.; Bungener, P.; Contat, F.; Montani, M.; Fuhrer, J. Grassland yield declined by a quarter in 5 years of free-air ozone fumigation. *Glob. Chang. Biol.* **2006**, *12*, 74–83. [CrossRef]

24. Cao, J.; Chen, Z.; Yu, H.; Shang, H. Differential responses in non-structural carbohydrates of *Machilus ichangensis* Rehd. et Wils. and *Taxus wallichiana* Zucc. Var. *chinensis* (Pilg.) florin seedlings to elevated Ozone. *Forests* **2017**, *8*, 323–334.

25. Chen, Z.; Cao, J.; Yu, H.; Shang, H. Effects of elevated ozone levels on photosynthesis, biomass and non-structural carbohydrates of *Phoebe bournei* and *Phoebe zhennan* in subtropical China. *Front. Plant Sci.* **2018**, *9*, 1764–1772. [CrossRef] [PubMed]

26. Agathokleous, E.; Saitanis, C.J.; Wang, X.; Watanabe, M.; Koike, T. A review study on past 40 years of research on effects of tropospheric O_3 on belowground structure, functioning, and processes of trees: A linkage with potential ecological implications. *Water Air Soil Pollut.* **2016**, *227*, 33. [CrossRef]

27. Feng, Z.; Li, P. Effects of ozone on Chinese trees. In *Air Pollution Impacts on Plants in East Asia*; Springer Japan: Tokyo, Japan, 2017; pp. 195–219.

28. Karlsson, P.E.; Uddling, J.; Skärby, L.; Wallin, G.; Sell-dén, G. Impact of ozone on the growth of birch (*Betula pendula*) saplings. *Environ. Pollut.* **2003**, *124*, 485–495. [CrossRef]

29. Oksanen, E.; Rousi, M. Differences of *Betula* origins in ozone sensitivity based on open-field experiment over two growing seasons. *Can. J. For. Res.* **2001**, *31*, 804–811. [CrossRef]

30. Winkler, B.; Fleischmann, F.; Gayler, S.; Scherb, H.; Matyssek, R.; Grams, T.E.E. Do chronic aboveground O_3 exposure and belowground pathogen stress affect growth and belowground biomass partitioning of juvenile beech trees (*Fagus sylvatica* L.)? *Plant Soil* **2009**, *323*, 31–44. [CrossRef]

31. Ainsworth, E.A. Understanding and improving global crop response to ozone pollution. *Plant J.* **2017**, *90*, 886–897. [CrossRef] [PubMed]

32. Li, P.; DeMarco, A.; Feng, Z.Z.; Anav, A.; Zhou, D.J.; Paoletti, E. Nationwide ground-level ozone measurements in China suggest serious risks to forests. *Environ. Pollut.* **2018**, *237*, 803–813. [CrossRef]

33. King, J.S.; Liu, L.; Aspinwall, M.J. Tree and forest responses to interacting elevated atmospheric CO_2 and tropospheric O_3: A synthesis of experimental evidence. In *Climate Change, Air Pollution and Global Challenges: Understanding Solutions from Forest Research*; Matyssek, R., Clarke, N., Cudlin, P., Mikkelsen, T.N., Tuovinen, J.-P., Wieser, G., Paoletti, E., Eds.; Elsevier Physical Sciences Series; Elsevier: San Diego, CA, USA, 2013; Volume 13, pp. 179–208.

34. Feng, Z.Z.; Tang, H.Y.; Uddling, J.; Pleijel, H.; Kobayashi, K.; Zhu, J.G.; Oue, H.; Guo, W.S. A stomatal ozone flux-response relationship to assess ozone-induced yield loss of winter wheat in subtropical China. *Environ. Pollut.* **2012**, *164*, 16–23. [CrossRef]

35. Hoshika, Y.; Katata, G.; Deushi, M.; Watanabe, M.; Koike, T.; Paoletti, E. Ozone-induced stomatal sluggishness changes carbon and water balance of temperate deciduous forests. *Sci. Rep.* **2015**, *5*, 9871. [CrossRef] [PubMed]

36. Bader, M.K.F.; Leuzinger, S.; Sonja, G.; Keel, S.G.; Rolf, T.W.; Siegwolf, R.T.W.; Hagedorn, F.; Schleppi, P.; Körner, C. Central European hardwood trees in a high-CO_2 future: Synthesis of an 8-year forest canopy CO_2 enrichment project. *J. Ecol.* **2013**, *101*, 1509–1519. [CrossRef]

37. Knepp, R.G.; Hamilton, J.G.; Mohan, J.E.; Zangerl, A.R.; Berenbaum, M.R.; DeLucia, E.H. Elevated CO_2 reduces leaf damage by insect herbivores in a forest community. *New Phytol.* **2005**, *167*, 207–218. [CrossRef] [PubMed]

38. Koike, T.; Watanabe, M.; Watanabe, Y.; Agathokleous, E.; Eguchi, N.; Takagi, K.; Satoh, F.; Kitaoka, S.; Funada, R. Ecophysiology of deciduous trees native to northeast Asia grown under FACE (Free Air CO_2 Enrichment). *J. Agric. Meteorol.* **2015**, *71*, 174–184. [CrossRef]

39. Norby, R.J.; Zak, D.R. Ecological lessons from free-air CO_2 enrichment (FACE) experiments. *Annu. Rev. Ecol. Evol. Syst.* **2011**, *42*, 181–203. [CrossRef]

40. Matyssek, R.; Kozovits, A.R.; Wieser, G.; King, J.; Rennenberg, H. Woody-plant ecosystems under climate change and air pollution-response consistencies across zonobiomes? *Tree Physiol.* **2017**, *37*, 706–732. [CrossRef]

41. Hiraoka, Y.; Iki, T.; Nose, M.; Tobita, H.; Tazaki, K.; Watanabe, A.; Fujisawa, I.; Kitao, M. Species characteristics and intraspecific variation in growth and photosynthesis of *Cryptomeria japonica* under elevated O_3 and CO_2. *Tree Physiol.* **2017**, *37*, 1–11. [CrossRef]

42. Hoshika, Y.; Watanabe, M.; Inada, N.; Koike, T. Ozone-induced stomatal sluggishness develops progressively in Siebold's beech (*Fagus crenata*). *Environ. Pollut.* **2012**, *166*, 152–156. [CrossRef]

43. Koike, T.; Kawaguchi, K.; Hoshika, Y.; Kita, K.; Watanabe, M.; Mao, Q.; Inada, N. Growth and photosynthetic responses of cuttings of a hybrid larch (*Larix gmelinii* var. japonica x *L. kaempferi*) to elevated ozone and/or carbon dioxide. *Asian J. Atmos. Environ.* **2012**, *6*, 104–110.

44. Kitao, M.; Tobita, H.; Kitaoka, S.; Harayama, H.; Yazaki, K.; Komatsu, M.; Agathokleous, E.; Koike, T. Light Energy Partitioning under Various Environmental Stresses Combined with Elevated CO_2 in Three Deciduous Broadleaf Tree Species in Japan. *Climate* **2019**, *7*, 79. [CrossRef]

45. Kitao, M.; Komatsu, M.; Yazaki, K.; Kitaoka, S.; Tobita, H. Growth overcompensation against O_3 exposure in two Japanese oak species, *Quercus mongolica* var. *crispula* and *Quercus serrata*, grown under elevated CO_2. *Environ. Pollut.* **2015**, *206*, 133–141. [PubMed]

46. Watanabe, M.; Yamaguchi, M.U.; Koike, T.; Izuta, T. Effects of ozone on Japanese trees. In *Air Pollution Impacts on Plants in East Asia*; Springer Japan: Tokyo, Japan, 2017; pp. 73–100.

47. Yamaguchi, M.; Watanabe, M.; Matsumura, H.; Kohno, Y.; Izuta, T. Experimental studies on the effects of ozone on growth and photosynthetic activity of Japanese forest tree species. *Asian J. Atmos. Environ.* **2011**, *5*, 65–78. [CrossRef]

48. Nakashizuka, T.; Iida, S. Composition, dynamics and disturbance regime of temperate deciduous forests in Monsoon Asia. *Vegetation* **1995**, *121*, 23–30. [CrossRef]

49. Kitao, M.; Yasuda, Y.; Kominami, Y.; Yamanoi, K.; Komatsu, M.; Miyama, T.; Mizoguchi, Y.; Kitaoka, S.; Yazaki, K.; Tobita, H.; et al. Increased phytotoxic O_3 dose accelerates autumn senescence in an O_3-sensitive beech forest even under the present-level O_3. *Sci. Rep.* **2016**, *6*, 32549. [CrossRef]

50. Watanabe, M.; Yamaguchi, M.U.; Koike, T.; Izuta, T. Growth and photosynthetic response of *Fagus crenata* seedlings to ozone and/or elevated carbon dioxide. *Landsc. Ecol. Eng.* **2010**, *6*, 181–190. [CrossRef]

51. Matsumura, H.; Mikami, C.; Sakai, Y.; Murayama, K.; Izuta, T.; Yonekura, T.; Miwa, M.; Kohno, Y. Impact of elevated O_3 and/or CO_2 on growth of *Betula platyphylla*, *Betula ermanii*, *Fagus crenata*, *Pinus densiflora* and *Cryptomeria japonica* seedlings. *J. Agric. Meteorol.* **2005**, *60*, 1121–1124. [CrossRef]

52. Watanabe, M.; Hoshika, Y.; Koike, T.; Izuta, T. Combined effects of ozone and other environmental factors on Japanese trees. In *Air Pollution Impacts on Plants in East Asia*; Springer Japan: Tokyo, Japan, 2017; pp. 101–110.

53. Kikuzawa, K. Leaf survival of woody plant in deciduous broad-leaved forests. 1. Tall trees. *Can. J. Bot.* **1983**, *61*, 2133–2139. [CrossRef]

54. Erbs, M.; Fangmeier, A. A chamberless field exposure system for ozone enrichment of short vegetation. *Environ. Pollut.* **2005**, *133*, 91–102. [CrossRef]

55. Farquhar, G.D.; von Caemmerer, S.; Berry, J.A. A biochemical model of photosynthetic acclimation in leaves of C_3 species. *Planta* **1980**, *149*, 78–90. [CrossRef]

56. Bernacchi, C.J.; Singsaas, E.L.; Pimentel, C.; Portis, A.R.; Long, S.P. Improved temperature response functions for models of Rubisco-limited photosynthesis. *Plant Cell Environ.* **2001**, *24*, 253–259. [CrossRef]

57. Schreiber, U.; Bilger, W.; Neubauer, C. Chlorophyll fluorescence as a nonintrusive indicator for rapid assessment of in vivo photosynthesis. In *Ecophysiology of Photosynthesis*; Schulze, E.-D., Caldwell, M.M., Eds.; Springer: Berlin/Heidelberg, Germany, 1994; pp. 49–70.

58. Sokal, R.R.; Rohlf, F.J. *Biometry*, 4th ed.; WH Freeman & Co: New York, NY, USA, 2011; p. 960.

59. Atkin, O.K.; Bloomfield, K.J.; Reich, P.B.; Tjoelker, M.G.; Asner, G.P.; Bonal, D.; Bönisch, G.; Bradford, M.G.; Cernusak, L.A.; Cosio, E.G.; et al. Global variability in leaf respiration in relation to climate, plant functional types and leaf traits. *New Phytol.* **2015**, *206*, 614–636. [CrossRef] [PubMed]

60. Warton, D.I.; Duursma, R.A.; Falster, D.S.; Taskinen, S. smart 3- an R package for estimation an inference about allometric lines. *Methods Ecol. Evol.* **2012**, *3*, 257–259. [CrossRef]

61. R Core Team. *R: A Language and Environment for Statistical Computing*; R Foundation for Statistical Computing: Vienna, Austria, 2015; Available online: http://www.R-project.org/ (accessed on 3 March 2016).

62. Agathokleous, E.; Watanabe, M.; Eguchi, N.; Nakaji, T.; Satoh, F.; Koike, T. Root production of *Fagus crenata* blume saplings grown in two soils and exposed to elevated CO_2 concentration: An 11-year free-air-CO_2 enrichment (FACE) experiment in northern Japan. *Water Air Soil Pollut.* **2016**, *227*, 187. [CrossRef]

63. Tobita, H.; Uemura, A.; Kitao, M.; Kitaoka, S.; Maruyama, Y.; Utsugi, H. Effects of elevated atmospheric carbon dioxide, soil nutrients and water conditions on photosynthetic and growth responses of *Alnus hirsuta*. *Funct. Plant Biol.* **2011**, *38*, 702–710. [CrossRef]

64. Karonen, M.; Ossipov, V.; Ossipova, S.; Kapari, L.; Loponen, J.; Matsumura, H.; Kohno, Y.; Mikami, C.; Sakai, Y.; Izuta, T.; et al. Effects of elevated carbon dioxide and ozone on foliar proanthocyanidins in *Betula platyphylla*, *Betula ermanii*, and *Fagus crenata* seedlings. *J. Chem. Ecol.* **2006**, *32*, 1445–1458. [CrossRef] [PubMed]

65. Landolt, W.; Buhlmann, U.; Bleuler, P.; Bucher, J.B. Ozone exposure-response relationships for biomass and root/shoot ratio of beech (*Fagus sylvatica*), ash (*Fraxinus excelsior*), Norway spruce (*Picea abies*) and Scot pine (*Pinus sylvestris*). *Environ. Pollut.* **2000**, *109*, 473–478. [CrossRef]

66. Matyssek, R.; Bytnerowicz, A.; Karlsson, P.E.; Paoletti, E.; Sanz, M.; Schaub, M.; Wieser, G. Promoting the O_3 flux concept for European forest trees. *Environ. Pollut.* **2007**, *146*, 587–607. [CrossRef]

67. Nikolova, P.S.; Andersen, C.P.; Blaschke, H.; Matyssek, R.; Häberle, K.H. Belowground effects of enhanced tropospheric ozone and drought in a beech/spruce forest (*Fagus sylvatica* L./*Picea abies* [L.] Karst). *Environ. Pollut.* **2010**, *158*, 1071–1078. [CrossRef]

68. Anderson, C.P.; Ritter, W.; Gregg, J.; Matyssek, R.; Grams, T.E.E. Below-ground carbon allocation in mature beech and spruce trees following long-term, experimentally enhanced O_3 exposure in Southern Germany. *Environ. Pollut.* **2010**, *158*, 2604–2609. [CrossRef]

69. Kinose, Y.; Fukamachi, Y.; Okabe, S.; Hiroshima, H.; Watanabe, M.; Izuta, T. Nutrient supply to soil offsets the ozone-induced growth reduction in *Fagus crenata* seedlings. *Trees* **2017**, *31*, 259–272. [CrossRef]

70. Watanabea, M.; Okabea, S.; Kinosea, Y.; Hiroshimaa, H.; Izuta, T. Effects of ozone on soil respiration rate of Siebold's beech seedlings grown under different soil nutrient conditions. *J. Agric. Meteorol.* **2019**, *75*, 39–46. [CrossRef]

71. Kolb, T.E.; Matyssek, R. Limitation and perspectives about scaling ozone impacts in trees. *Environ. Pollut.* **2001**, *115*, 373–393. [CrossRef]

72. Watanabe, M.; Hoshika, Y.; Koike, T. Photosynthetic responses of Monarch birch seedlings to differing timing of free air ozone fumigation. *J. Plant Res.* **2014**, *127*, 339–345. [CrossRef] [PubMed]

Article

Grapevine and Ozone: Uptake and Effects

Ivano Fumagalli [1], Stanislaw Cieslik [1,*], Alessandra De Marco [2], Chiara Proietti [3] and Elena Paoletti [4]

[1] European Commission, Joint Research Centre, via E. Fermi 2749, I-21027 Ispra (VA), Italy; Ivano.FUMAGALLI@ec.europa.eu

[2] Italian National Agency for New Technologies, Energy and the Environment (ENEA), C.R. Casaccia, via Anguillarese 301, I-00123 S. Maria di Galeria, Italy; alessandra.demarco@enea.it

[3] Italian Institute for Environmental Protection and Research (ISPRA), via Brancati 48, I-00144 Rome, Italy; chiara.proietti@isprambiente.it

[4] Institute of Research on Terrestrial Ecosystems (IRET), National Research Council (CNR), via Madonna del Piano 10, I-50019 Sesto Fiorentino, Italy; elena.paoletti@cnr.it

* Correspondence: stanislaw.cieslik@yahoo.it

Received: 19 October 2019; Accepted: 10 December 2019; Published: 12 December 2019

Abstract: The grapevine (*Vitis vinifera*, L.) has been long since recognized as an ozone-sensitive plant. Ozone molecules can penetrate grapevine leaf tissues when the concentration of ozone in the atmosphere is high due to air pollution. This causes cell damage and interferes with photosynthetic mechanisms, subsequently slowing down plant growth and resulting in premature leaf senescence. Secondary effects include changes in biochemical processes that affect the chemical composition of the must and are likely to alter the quality of the wine. An experiment was conducted during two grapevine-growing seasons in 2010 and 2011 to gain knowledge of the effect of high ozone levels on the yield and on several biochemical characteristics of the plant which could influence the quality of the final product. These factors are economically important for agricultural production; this is especially true for Italy, which is one of the largest wine producers worldwide. The method used was a facility consisting of open top chambers operated at a vineyard in Angera (northern Italy). This facility permitted the study of the effects of different ozone levels. At the end of the experiment, the grapes were weighed and chemical analyses were carried out in order to understand the effects of ozone on the different characteristics of the grapes and on concentrations of several of its chemical substances. In particular, concentrations of yeast assimilable nitrogen, degrees Brix, pH, tartaric and malic acids, and polyphenols, including resveratrol, were considered, as these influence the quality of the wine. Parameters characterizing the different ozone levels were expressed in terms of ozone exposure (AOT40) and phytotoxic ozone dose (POD). The results showed that high ozone levels affect grapevine weight and thus its yield. In addition, the quality of the wine is affected by reductions of polyphenols which diminish the nutritional benefits of the product. In addition, these reductions cause the wine to have a more aggressive taste. These results emphasize the practical importance of the present study.

Keywords: ozone uptake; *Vitis vinifera*; open top chambers; ozone damage metrics; wine quality

---

1. Introduction

Sensitivity of the grapevine (*Vitis vinifera* L.) to ozone has been known for a rather long time [1–5]; other authors [6,7] reported visible foliar injury in plants of *V. vinifera* grown on vineyards in Greece, Spain, and Italy. When the atmospheric concentration of ozone (O_3) is high due to air pollution, ozone molecules penetrate grapevine leaf tissues through the stomata, causing cell damage and interfering with photosynthetic mechanisms, something which produces a subsequent slowing down of plant growth and premature leaf senescence in many plant species [6,8,9]. Secondary effects include changes in biochemical processes that affect the chemical composition of the must and which are likely to cause alterations in the quality of the wine produced [10].

The effects of ozone uptake by grapevines are part of the much wider problem of ozone damage to crops, which has been known about for a long time [11–13]. Reviews on this topic can be found in e.g., [14,15]. The latter author considers ozone to be the most important air pollutant in non-urban areas and concludes that it considerably affects ecosystems, whether natural or cultivated.

The way by which ozone damages vegetation is related to its oxidizing potential. After penetration through the stomata, ozone reaches the cell membranes and the apoplast, whose oxidation produces hydroxyl and peroxyradicals as well as hydrogen peroxide [16]. Alkenes are converted into aldehydes and peroxides [17]. All these oxidized molecules are known under the generic appellation of reactive oxygen species (ROS). Details on these mechanisms can be found in [18,19]. Cell death may occur as the ultimate consequence of these oxidation processes. Up to a certain point, detoxification processes, involving ascorbate and glutathione [20], can "clean" the cell from ozone attack. The detoxification capability is genetically controlled, meaning sensitivity of plants to ozone is strongly species- and even cultivar-dependent. The final result of the action of ozone on vegetation is the perturbation of photosynthesis, which leads in several cases to a reduction in growth rates. This can cause important losses of agricultural crop yields on a global scale [21,22], with consequences at economic and human levels. The uptake of ozone by vegetation also causes physiological perturbations to natural vegetation [23]. Other studies have analyzed the effects of ozone on natural biomass production [24–27].

Two approaches are usually followed for the assessment of O_3 risk to crops: the concentration-based approach [28–30] and the flux-based approach [31]. The former estimates ozone risk to vegetation by calculating an exposure index named AOT40, which is defined as the accumulated hourly O_3 concentration exceeding a threshold of 40 ppb that is summed up over a growing season. The latter approach is based on another index, the phytotoxic ozone dose with threshold y (PODy) (mmol.m^{-2}), which is computed as the accumulated stomatal ozone flux and is similarly summed up over a growing season. The latter depends on plant phenology, solar irradiance, temperature, vapor pressure deficit, and soil moisture [32,33]. A critical review of these indices has been made by other authors [34].

To date, the metrics described above have been applied to *V. vinifera* in a few studies only [5,10,35]. In the latter work cited, the authors examined the possible effects of O_3 on grape leaves under natural conditions in the Rijeka bay area (Croatia) using the AOT40 index. The effects of environmental factors associated with air pollution on grapevines have been investigated by [29,36], among others. These authors found adverse effects of high temperature and air pollution conditions on photosynthesis in *Vitis vinifera* (*cv*. Hongti). A reduced value of stomatal conductance and of net assimilation in two grapevine cultivars (Maturano and San Giuseppe) while these were fumigated with O_3 was also found [37]. Furthermore, [10] studied the impact of stomatal uptake of ozone on the quality of must as well as fruit yield in pot-grown grapevine plants exposed to different O_3 concentration levels. These authors found foliar damage as well as yield losses. These results were obtained in controlled experimental conditions that were not necessarily representative of grapevines growing on a vineyard.

The goal of the experiment described here was to study the possible effects that ozone uptake by grapevines could cause in natural conditions and when the plants are enclosed in open top chambers (OTCs). Determination of the chemical composition of the must as well as of the weight of the grapes, expressing their biomass production, was carried out. Inside the OTCs, the O_3 concentration was controlled and reduced by filters. The campaigns were conducted in August-September 2010 and in May-September 2011 at a vineyard near the town of Angera, which is located in northern Italy.

In the following sections the experiment, including the measurement of ozone concentration and other relevant parameters, is described. Then, the determination of exposure and integrated ozone flux is explained, as well as the measurable effects of ozone on yields and on chemical contents of the must.

2. Materials and Methods

Four OTCs were operated on our chosen site (see Figure 1). Each OTC enclosed four grapevine plants (*Vitis vinifera* L., cv. Merlot). The size of each OTC was 0.8 m × 4.0 m × 2.5 m. Two OTCs were fed with ambient air and the two others with filtered air, from which a part of the ambient O_3 was removed by a chemical substance (PURAFIL$^{®}$, Doraville, GA, USA). It was deduced from the measurement of flow rates of the air pushed into the chambers (16 m^3 min^{-1}) and from the volume of the OTCs that air inside the OTCs was completely mixed and renewed in two minutes. Measurement of temperature and humidity at various points inside the OTCs led to the conclusion that these variables were spatially homogenous. Ozone concentrations were monitored inside and outside the chambers (both filtered and unfiltered OTCs) during the whole experiment by making use of an ozone analyzer (Thermo Environmental Instruments, Inc., mod. 49C, Waltham, MA, USA). The analyzer's input line was connected with a three-way switching valve permitting alternation between three input lines, pumping air from the open air as well as from the filtered and unfiltered OTCs.

Figure 1. View of two of the four open top chambers enclosing grapevine plants on the chosen vineyard in Angera. The chambers were fed with air which was ambient or charcoal-filtered.

The first phase of the experiment was merely aimed at testing the capability of the experimental setup. This phase took place during the summer of 2010. During the initial period (5–11 August), the system was operated without eliminating ozone by filtering. The ozone concentrations recorded for the three lines (i.e., two pairs of OTCs and the external open air) are represented in Figure 2, and show good agreement between the lines. For the period covering 15 August to harvest time, filtering was introduced using PURAFIL$^{®}$-filled filters, leading to the conclusion that absorption of ozone within the filters caused an around 30% reduction in the concentrations (Figure 3). An additional aspect to consider is the difference between open-air O_3 concentrations and those measured inside the unfiltered chambers. This difference is related to a loss of ozone in the unfiltered OTCs due to its destruction on the tube walls.

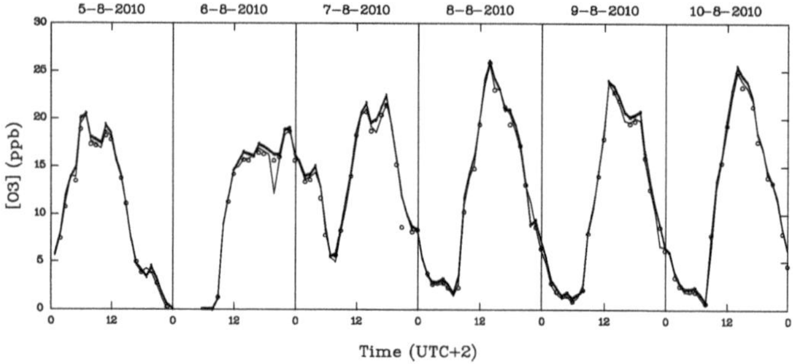

Figure 2. Ozone concentrations measured inside and outside the open top chambers enclosing grapevines at the vineyard in Angera. Measurements were carried out from 5 to 10 August 2010 without any filtering in order to check whether the outputs obtained with the three lines agreed with each other. No significant difference appeared.

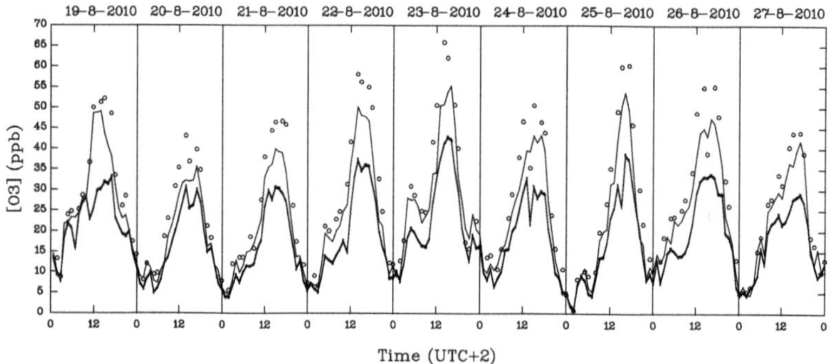

Figure 3. Ozone concentrations measured inside and outside the open top chambers (OTCs) during August 2010 at the vineyard in Angera. Here, filtering was active. The bold line refers to the filtered OTC, the thin line to the unfiltered OTC, and the circles to the open air.

During the year 2011, the complete growing season of grapevines was investigated. The experiment took place from June 2011 to harvest time, which occurred in September. In two chambers the air was filtered in order to reduce the O_3 content. In each OTC, four plants were present. Since for each treatment (filtered and unfiltered) two OTCs were set up, each treatment consisted of eight plants. We also sampled eight plants for the open-air condition. Sampling was performed by randomly marking two bunches per plant at the beginning of the experiment (June 2011). These bunches were cut at harvest time (in September). However, at harvest, we had to eliminate a few plants because of damage by parasites or other causes like branches growing outside the OTCs. In the end, sampling included seven plants for the filtered OTCs, six plants for the unfiltered OTCs, and six plants for the external open-air plants. Bunches cut were weighed and then sent to the Istituto Agrario di San Michele all'Adige (Edmund Mach Foundation) for chemical analysis.

Chemical characteristics such as degrees Brix (an index quantifying the dissolved saccharose content of the must), pH, concentrations of tartaric and malic acids, potassium, and assimilable nitrogen were determined by FTIR. Total anthocyanins and polyphenols were determined using a method described by Mattivi [38]; resveratrol was determined by high performance liquid chromatography with a diode array detector [39].

ANOVA was used to statistically analyze the results obtained (see the Results section below).

3. Results

The main goal of the analysis was to determine whether grapes exposed to ozone-rich air (unfiltered OTC air and open air, i.e., air external to the OTCs) showed differences from grapes exposed to ozone-poor air (filtered OTC air) with regard to various parameters such as yield and chemical composition data.

The experiments started in the beginning of June, during the summer of 2011. Ozone concentration, temperature, and relative humidity were monitored continuously until September when harvesting took place. The average temperature was 22.6 °C in the open air (OA), 23.3 °C in the filtered OTCs (F) and 22.7 °C in the unfiltered OTCs (NF). This indicates that the temperature differences were not critically important. As far as relative humidity (RH) was concerned, differences were higher between the OA and the air inside both kinds of OTCs. In the OA, the average RH was 78%, whereas in F and NF it was 76%. A more complete picture of temperature and humidity is shown in Figure 4. In parallel to the monitoring of concentrations, ozone stomatal fluxes were calculated by using the multiplicative DO_3SE model (Deposition of Ozone for Stomatal Exchange) which was developed by several authors recently [32,33]. An example of a recorded time series is shown in Figure 5, where O_3 concentration inside and outside the OTCs is represented, as well as temperature, relative humidity, and calculated stomatal O_3 flux, for the period 25–31 July 2011.

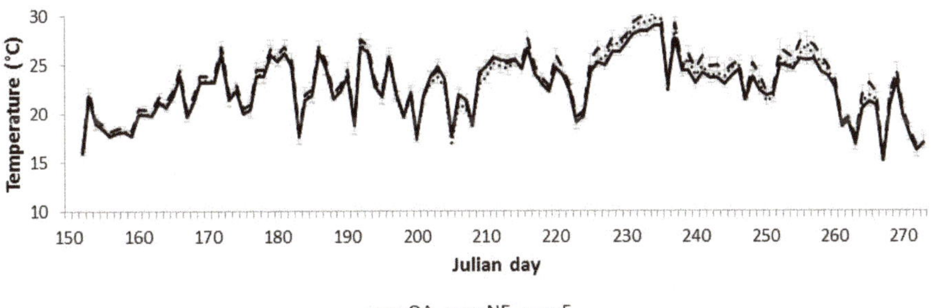

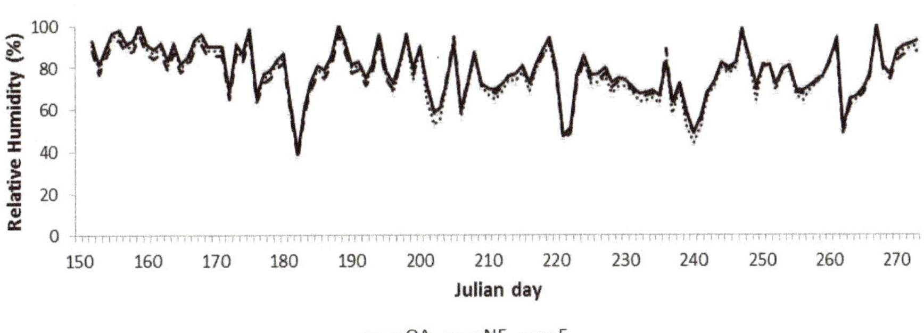

Figure 4. Time series of daily averaged temperature (upper frame) and relative humidity (lower frame) during the 2011 experiment. No significant differences between the three cases (open air (OA), filtered OTC (F) and unfiltered OTC (NF)) can be noted.

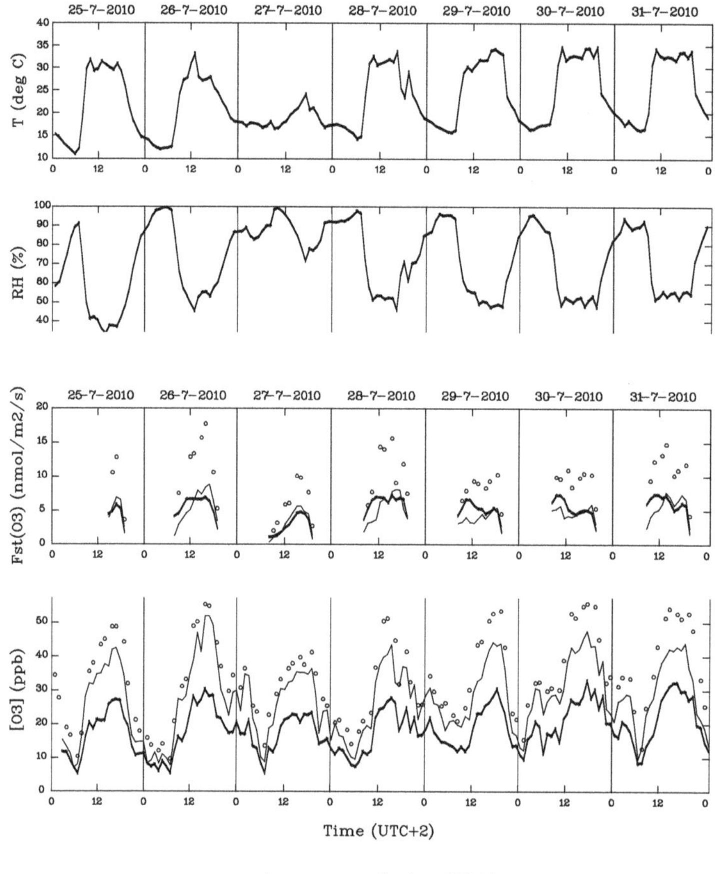

Figure 5. Time series of data recorded at the Angera OTC site from 25 to 31 July 2011. The rows from top to bottom are for ambient temperature, ambient relative humidity, stomatal ozone fluxes calculated using the DO$_3$SE model (Deposition of Ozone for Stomatal Exchange), and ozone concentrations. For the two lowest rows, the circles represent data for outside the OTCs, the thin solid line corresponds to data for the unfiltered OTCs, and the bold line corresponds to data for the filtered OTCs.

Average trends in O$_3$ concentration, flux, and temperature during the growing season of 2011 are shown in Figure 6. To facilitate the interpretation of these time series, the figure shows daily averages of the variables. A rough correlation appears between ozone concentration and temperature. This reflects the mechanisms of ozone formation, mainly due to photochemical processes, as the mechanisms are triggered by intense solar radiation. Periods of high solar radiation generally occur during summer when temperatures are higher. Stomatal ozone fluxes tend instead to show lower values when the temperature is higher, and, more generally, ozone fluxes and concentrations are not necessarily correlated. High temperatures usually cause stomatal closure, and the daily cycles of these two variables follow different patterns.

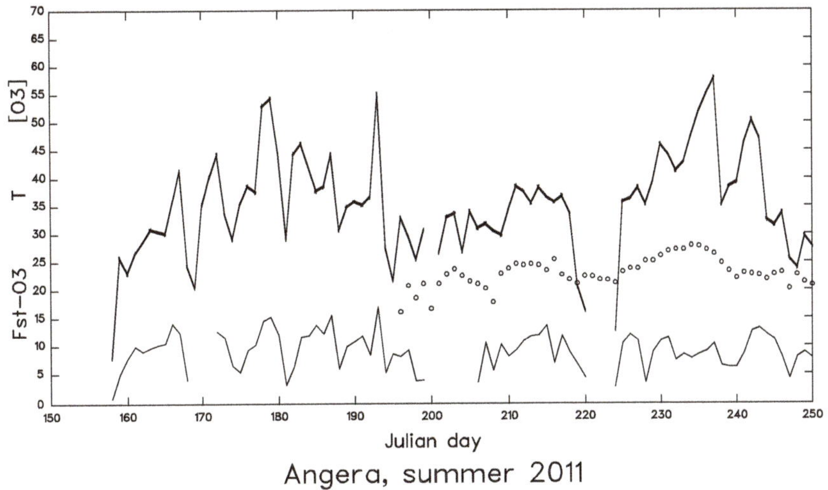

Figure 6. Daily averaged ozone concentrations ($[O_3]$) (ppb, bold line), open air temperatures (T) (°C, circles), and stomatal ozone fluxes (Fst-O_3) (nmol m^2 s^{-1}, thin solid line) at the Angera vineyard from June to September 2011.

By integrating the stomatal fluxes over time from anthesis to harvest, we determined the phytotoxic ozone dose (POD), which expresses the total amount of O_3 taken up by the plant (the dose) per unit area across the complete growing season.

Since the recorded O_3 concentration time series was practically uninterrupted from anthesis to harvest, it was possible to determine the exposure indices (AOT40) [30] for the three inlet lines of the ozone measuring system for open air, non-filtered, and filtered OTCs. These values are reported in Table 1. The cumulative curves corresponding to these calculations are shown in Figure 7. A similar calculation was carried out for the stomatal ozone fluxes, which were calculated using the DO$_3$SE model. Here, two thresholds were used: 0 (POD$_0$) and 3 (POD$_3$). Cumulative curves are also represented in Figure 7 and the POD$_0$ values are reported in Table 2.

After harvest, the grapes were analyzed at the Edmund Mach Foundation. Results of these analyses are shown in Table 2, where the results, expressed as averages per treatment, i.e., plants growing externally to the OTCs in open air, in filtered OTCs, and in unfiltered OTCs, respectively, are shown in three columns, with each column corresponding to each treatment.

Table 1. Values of the AOT40 and POD$_0$ indices calculated for summer 2011.

	AOT40 (ppb.h)	POD$_0$ (mmol.m^{-2})
External	12,300	38.1
Unfiltered OTC	6100	32.0
Filtered OTC	2000	21.4

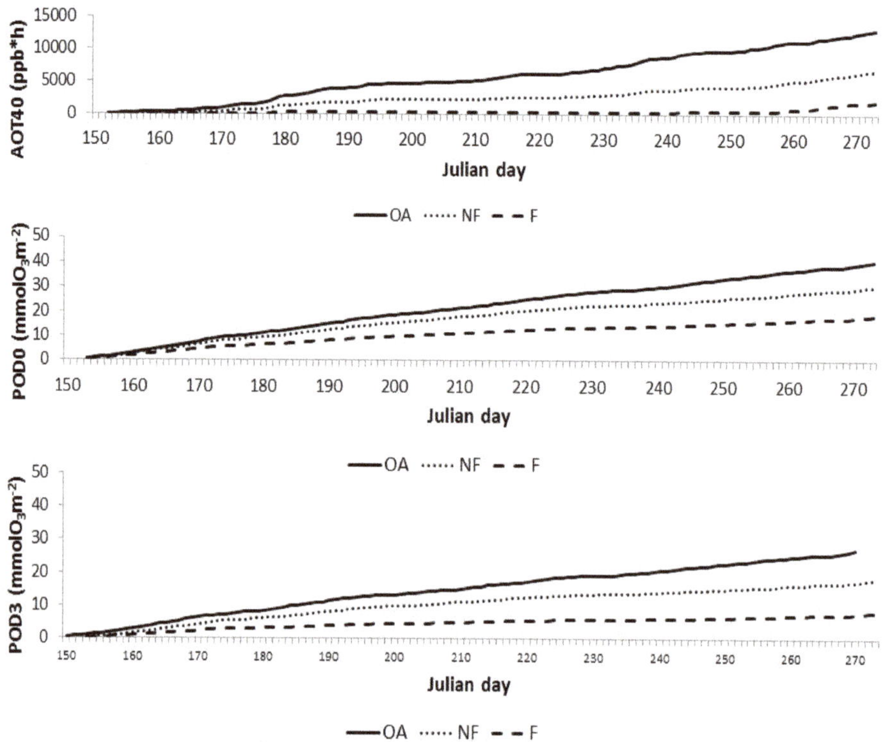

Figure 7. Ozone exposure index (AOT40) (upper frame), phytotoxic ozone dose without threshold (POD_0) (middle frame), and phytotoxic ozone dose with threshold (POD_3) (lower frame); the data are presented as progressively accumulated over the period of the experiment under the three experimental conditions.

In order to determine the effect of ozone uptake on grapevines, ANOVA was performed on nine out of the fourteen variables appearing in Table 2. We also used the Tukey post-hoc test for multiple comparisons between pairs of treatments. The results of this procedure have been visualized by the box plots shown in Figure 8. Each graph shows, for the variable under consideration, three rectangles which correspond to the three treatments: filtered OTCs, unfiltered OTCs, and external treatment (open air). The horizontal bar present in each rectangle represents the median value (the 50th percentile, i.e., not the mean value, which is shown in Table 2). The vertical extensions of the rectangles represent the lower and upper quartiles (25th and 75th percentiles, respectively). The vertical bars crossing the rectangles represent the minimum and maximum values of the samples. For each variable, the p values were calculated for the three treatments together, as well for pairs of two treatments (filtered-external, unfiltered-external, and unfiltered-filtered), in order to detect significant differences, using the Tukey test. If the p values were less than 0.05, differences were considered significant. We now examine how each of these variables reacted.

Table 2. Results of analyses made on Angera grapevine bunches for summer 2011. The values are averages over all bunches cut from plants corresponding to the different treatments; the standard deviations (sigma) are also represented. Analyses were carried out at the Edmund Mach Foundation.

	External	Filtered OTC	Unfiltered OTC
Grape weight per bunch (kg)	2.14 ± 0.45	2.50 ± 0.37	1.72 ± 0.54
Degrees Brix (°Bx)	20.62 ± 0.35	21.18 ± 0.42	20.94 ± 0.11
pH	3.32 ± 0.0	3.32 ± 0.1	3.35 ± 0.0
Titrable acidity (as tartaric acid, g/L)	5.35 ± 0.1	5.56 ± 0.2	5.37 ± 0.1
Density at 20 °C	1.090 ± 0.0	1.092 ± 0.0	1.091 ± 0.0
Tartaric acid (g/L)	6.25 ± 0.1	6.19 ± 0.3	6.15 ± 0.1
Malic acid (g/L)	2.41 ± 0.12	2.65 ± 0.40	2.40 ± 0.10
Potassium (mg/L)	1705 ± 27.9	1745 ± 69.8	1689 ± 37.2
Assimilable nitrogen (mg/L)	103.5 ± 14.3	76.0 ± 8.6	87.1 ± 10.0
Total anthocyanins (mg/kg)	569.3 ± 131.3	731.0 ± 98.8	668.7 ± 62.8
Total polyphenols (mg/kg)	992 ± 103	1141 ± 53	972 ± 22
Trans-resveratrol (mg/kg)	2.5 ± 0.8	3.1 ± 0.7	3.6 ± 1.1
Cis-resveratrol glucoside (mg/kg)	<0.1	<0.1	<0.1
Trans-resveratrol glucoside (mg/kg)	46.3 ± 6.7	40.4 ± 0.7	37.3 ± 3.7

1. Grape weight per bunch. There is a significant difference between the samples obtained from the filtered OTCs and from the other two treatments ($p = 0.0213$). However, the p value corresponding to the unfiltered versus external couple is too high (0.22) to imply any significance. Notwithstanding the difference in ozone levels between the unfiltered and external treatments, little difference in the bunch weights is detectable. This is probably due to an effect of the open top chambers in which the physical conditions were not the same as in the open air (i.e., for the external samples). However, significant differences between the weight of the bunches enclosed in the filtered and unfiltered OTCs are clearly visible (see Figure 8), with $p = 0.0166$. This result shows that higher ozone levels cause a reduction in yields, at least in the case of our experiment.

2. Degrees Brix. This variable expresses the sugar content of the must and is widely used in the wine production industry. A rather important effect of ozone levels appears (see Figure 8), especially between samples enclosed in filtered OTCs and external samples ($p = 0.0022$). The difference is less clear between filtered and unfiltered samples.

3. Polyphenols. Here, no clear difference appears between unfiltered and external samples, but the samples enclosed in filtered OTCs are characterized by a much higher content of polyphenols ($p = 0.045$). These substances are known as antioxidants and have health effects on digestion, for example. The results presented here indicate that high ozone content in air can counterbalance the health effects of polyphenols present in wine.

4. Anthocyanins. These pigments present in grapes show a significant difference ($p = 0.046$) between plants enclosed in filtered OTCs and external ones.

5. Malic acid. No significant difference was found between the three treatments due to the overlap of variability ranges for the different treatments. All p values lie above 0.2.

6. Titrable acidity. Differences appear rather clearly between plants enclosed in filtered and unfiltered OTCs ($p = 0.0086$) and between those enclosed in filtered OTCs and under external conditions ($p = 0.0055$).

7. Assimilable nitrogen. Here, an opposite pattern with respect to most other variables can be seen: higher ozone levels result in higher assimilable nitrogen. The variance analysis for all

three treatments combined shows good significance ($p = 0.0032$); the highest significance appears between the external plants and those enclosed in filtered OTCs ($p = 0.00067$).

8. Trans-resveratrol. As for malic acid, no significant difference was found between the three treatments. All p values lie above 0.1.

9. Trans-resveratrol glucoside. A significant difference ($p = 0.033$) appears between external plants and those enclosed in unfiltered OTCs. However, for this variable, no conclusion can be drawn about the effect of ozone between filtered and unfiltered OTCs.

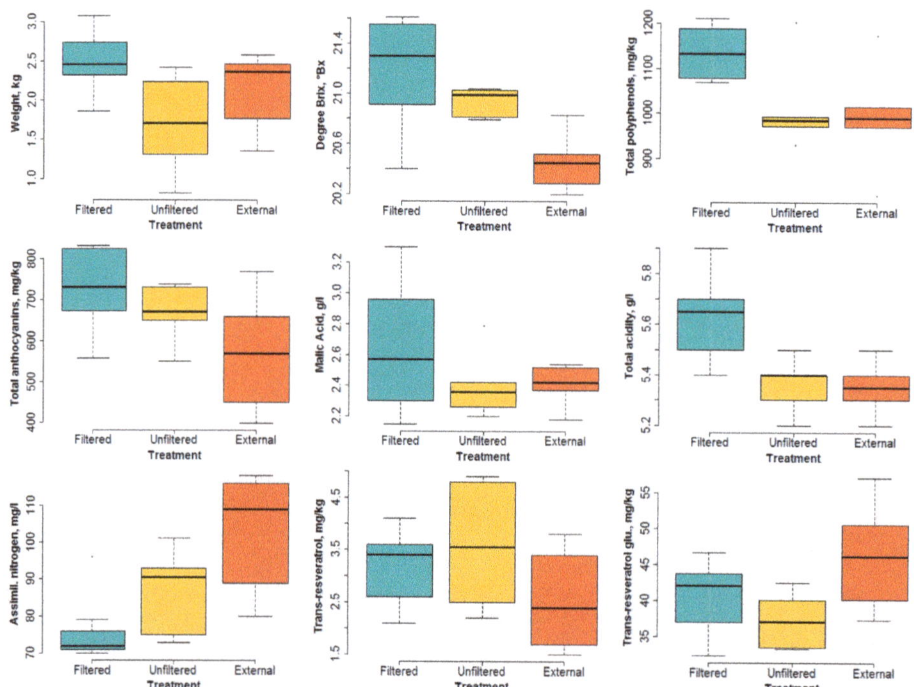

Figure 8. Box plots representing the behavior of nine variables characterizing the grapevines' responses when subject to three treatments: filtered and unfiltered OTCs and external (open air) conditions. The first variable is the weight per bunch; all other variables are chemical variables.

4. Discussion and Conclusions

The effect of atmospheric O_3 on *Vitis vinifera* has been known for a long time [2] as far as foliar injury and some other physiological parameters are concerned. No direct quantitative observations of parameters like crop yield and relevant chemical variables that influence the quantity and quality of the final product, i.e., wine, have been available to date. The present work aimed to constitute a step forward towards fill this knowledge gap by using the open top chamber technique, in which grapevine plants are grown under controlled O_3 concentrations. The results show that higher O_3 concentrations cause decreases in crop yield, expressed as the weights of grapevine bunches. In addition, the concentrations of several chemical substances present in the must, which influence the quality of the wine produced and are subject to changes in the case of high tropospheric ozone levels. These substances are polyphenols and saccharose (expressed as degrees Brix). Anthocyanins and total acidity are affected as well.

If we exclude the measurements made in 2010, which were just intended as a test of the facility, the present study was carried out over just one season of grapevine growth and was focused on a limited amount of plant material. Still, the results of the present study show that adverse effects can be observed when grapevine plants are exposed to high ozone levels. While this work can be considered as a step forward, further studies are needed for a more comprehensive picture. It was found that exposures to high O_3 concentrations reduced the quantities of grapes harvested, resulting in wine production losses. In addition, high O_3 levels also affected the quality of the final product. The polyphenol content was also reduced in the case of high O_3 levels. These substances have beneficial properties for human health. However, other parameters such as pH and density seemed unaffected. On examining Figure 8, the effect of O_3 uptake by vine plants still appears for several variables.

Owing to the economic importance of wine production and quality, it is recommended that full-scale experiments be conducted in particular by focusing on longer exposure periods and a higher amount of plant material. They would need to be coupled with direct measurements of ozone flux, e.g., by using eddy covariance techniques to directly measure ozone fluxes.

Author Contributions: I.F. and S.C. organised and carried out the field experiment; They also participated in the analysis of the results; E.P. made a conceptual work in designing the experiment; C.P. and A.D.M. participated in the modeling and analysis of the data.

Funding: No funding external to the participating institutes was made.

Acknowledgments: The authors are indebted to the Edmund Mach Institute, S. Michele all'Adige, Italy, for having carried out the chemical and gravimetrical analyses of the grapevine bunches harvested after the experiment. The authors thank Michel Gerboles for his valuable assistance in the statistical treatment of the results.

Conflicts of Interest: The authors declare no conflict of interest.

References

1. Popescu, S.M. The combined effects of CO_2 and O_3 on the physiology of the grapevine *Vitis vinifera* L. cv. Merlot). *J. Hortic. For. Biotechnol.* **2011**, *15*, 62–66.
2. Richards, B.L.; Middleton, J.T.; Hewitt, W.B. Ozone stipple of grape leaf lesions on the upper leaf surfaces and premature leaf fall occur on grapevines in areas polluted by air-borne ozone. *Calif. Agric.* **1959**, *13*, 4–11.
3. Roper, T.R.; Williams, L.E. Effects of ambient and acute partial pressures of ozone on leaf net CO_2 assimilation of field-grown *Vitis vinifera* L. *Plant Physiol.* **1989**, *91*, 1501–1506. [CrossRef] [PubMed]
4. Shertz, R.D.; Kender, W.D.; Musselman, R.D. Effects of ozone and sulfur dioxide on grapevines. *Sci. Hortic.* **1980**, *13*, 37–45. [CrossRef]
5. Soja, G.; Eid, M.; Gangl, H.; Redl, H. Ozone sensitivity of grapevine (*Vitis vinifera* L.): Evidence for a memory effect in a perennial crop plant. *Phyton* **1997**, *37*, 265–270.
6. Fumagalli, I.; Gimeno, B.S.; Velissariou, D. Evidence of ozone-induced adverse effects on crops in the Mediterranean region. *Atmos. Environ.* **2001**, *35*, 2583–2587. [CrossRef]
7. Velissariou, D.; Gimeno, B.S.; Badiani, M.; Fumagalli, I.; Davison, A.W. Records of O3 visible injury in the ECE Mediterranean region. In *Critical Levels for Ozone in Europe: Testing and Finalising the Concept*; Karelampi, L., Skarby, L., Eds.; UN-ECE Workshop Report; University of Kuopio: Kuopio, Finland, 1996; pp. 343–350.
8. Fuhrer, J.; Egger, A.; Lehnherr, B.; Grandjean, A.; Tschannen, W. Effects of ozone on the yield of spring wheat grown in open-top field chambers. *Environ. Pollut.* **1989**, *60*, 273–289. [CrossRef]
9. Oksanen, E.; Freiwald, V.; Prozherina, N.; Rousi, M. Photosynthesis of birch is sensitive to springtime frost and ozone. *Can. J. For. Res.* **2005**, *35*, 703–712. [CrossRef]
10. Soja, G.; Reichenauer, T.G.; Eid, M.; Soja, A.M.; Schaber, R.; Gangl, H. Long-term ozone exposure and ozone uptake of grapevines in open-top chambers. *Atmos. Environ.* **2004**, *38*, 2313–2321. [CrossRef]
11. Lea, M.C. On the influence of ozone and some other chemical agents on germination and vegetation. *Am. J. Sci. Arts* **1864**, *37*, 373–376. [CrossRef]
12. Rich, S.; Waggoner, P.E.; Tomlinson, H. Ozone uptake by bean leaves. *Science* **1970**, *169*, 79. [CrossRef] [PubMed]
13. Turner, N.C.; Waggoner, P.E.; Rich, S. Removal of ozone from the atmosphere by soil and vegetation. *Nature* **1974**, *250*, 486. [CrossRef]
14. Ashmore, M.R. Assessing the future global impacts of ozone on vegetation. *Plant Cell Environ.* **2005**, *28*, 949–964. [CrossRef]

15. Krupa, S.V.; Manning, W.J. Atmospheric ozone: Formation and effects on vegetation. *Environ. Pollut.* **1988**, *50*, 101–137. [CrossRef]

16. Grimes, H.D.; Perkins, K.K.; Boss, W.F. Ozone degrades into hydroxyl radical under physiological conditions. *Plant Physiol.* **1983**, *72*, 1016–1020. [CrossRef]

17. Pryor, W.A.; Church, D.F. Aldehydes, hydrogen peroxide and organic radicals, as mediators of ozone toxicity. *Free Rad. Biol. Med.* **1991**, *11*, 41–46. [CrossRef]

18. Byvoet, P.; Balis, J.U.; Shelley, S.A.; Montgomery, M.R.; Barrber, M.J. Detection of hydroxyl radicals upon interaction of ozone with aqueous media of extracellular surfactants. *Arch. Biochem. Biophys.* **1995**, *319*, 464–469. [CrossRef]

19. Pell, E.J.; Schangnhaufer, C.D.; Arteca, R.N. Ozone-induced oxidative stress: Mechanisms of action and reaction. *Physiol. Plant.* **1997**, *100*, 264–273. [CrossRef]

20. Anjum, N.A.; Umar, S.; Chan, M.T. (Eds.) *Ascorbate-Glutathione Pathway and Stress Tolerance in Plants*; Springer: Dordrecht, The Netherlands, 2010; ISBN 978-90-481-9403-2.

21. Tai, A.P.K.; San Martin, M. Impacts of ozone pollution and temperature extremes on crop yields: Spatial variability, adaptation and implications for future food security. *Atmos. Environ.* **2017**, *169*, 11–21. [CrossRef]

22. Van Dingenen, R.; Dentener, F.J.; Raes, F.; Krol, M.C.; Emberson, L.; Cofala, J. The global impact of ozone on agricultural crop yields under current and future air quality legislation. *Atmos. Environ.* **2009**, *43*, 604–618. [CrossRef]

23. Sitch, S.; Cox, P.M.; Collins, W.J.; Huntingford, C. Indirect forcing of climate change through ozone effects on the land-carbon sink. *Nature* **2007**, *448*, 791–794. [CrossRef] [PubMed]

24. Cieslik, S.; Tuovinen, J.P.; Baumgarten, M.; Matyssek, R.; Brito, P.; Wieser, G. Gaseous exchange between forests and the atmosphere. In *Climate Change, Air Pollution and Global Challenges*; Matyssek, R., Clarke, N., Cudlin, P., Mikkelsen, T.N., Tuovinen, J.P., Wieser, G., Paoletti, E., Eds.; Elsevier: Amsterdam, The Netherlands, 2013; part 2.

25. Franz, M.; Alonso, R.; Arneth, A.; Zaehle, S. Evaluation of simulated ozone effects in forest ecosystems against biomass damage estimates from fumigation experiments. *Biogeosciences* **2018**, *15*, 6941–6957. [CrossRef]

26. Mills, G.; Pleijel, H.; Malley, C.S.; Sinha, B.; Cooper, O.R.; Scholtz, C.S.; Neufeld, H.S.; Simpson, D.; Sharops, K.; Feng, Z.; et al. Tropospheric ozone assessment report: Present-day tropospheric ozone distribution and trends relevant to vegetation. *Elementa* **2018**, *6*, 47. [CrossRef]

27. Niinemets, Ü. Responses, of forest trees to single and multiple environmental stresses from seedlings to mature plants. *For. Ecol. Manag.* **2010**, *260*, 1623–1639. [CrossRef]

28. Directive 2008/50/EC of the European Parliament and of the Council. *Off. J. Eur. Communities* **2008**, *11*, 2008.

29. Grünhage, L.; Jäger, H.J. From critical levels to critical loads for ozone: A discussion of a new experimental and modelling approach for establishing flux-response relationships for agricultural crops and native plant species. *Environ. Pollut.* **2003**, *125*, 99–110. [CrossRef]

30. Kärenlampi, L.; Skärby, L. (Eds.) *Critical Levels for Ozone in Europe: Testing and Finalizing the Concepts*; UN-ECE Workshop Report; University of Kuopio: Kuopio, Finland, 1996.

31. Paoletti, E.; Manning, W.J. Toward a biologically significant and usable standard for ozone that will also protect plants. *Environ. Pollut.* **2007**, *150*, 85–95. [CrossRef]

32. Emberson, L.D.; Wieser, G.; Ashmore, M.R. Modelling of stomatal conductance and ozone flux of Norway spruce: Comparison with field data. *Environ. Pollut.* **2000**, *109*, 393–402. [CrossRef]

33. Emberson, L.D.; Ashmore, M.R.; Cambridge, H.M.; Simpson, D.; Tuovinen, J.P. Modelling of stomatal ozone flux across Europe. *Environ. Pollut.* **2000**, *109*, 403–413. [CrossRef]

34. Musselman, R.C.; Lefohn, A.S.; Massman, W.J.; Heath, R.L. A critical review and analysis of the use of exposure- and flux-based ozone indices for predicting vegetation effects. *Atmos. Environ.* **2006**, *40*, 1869–1888. [CrossRef]

35. Alebić-Juretić, A.; Bokan-Vocelić, I.; Mifka, B.; Zatezalo, M.; Zubak, V. Impact of ozone gradient on grapevine leaves. In Proceedings of the 19th EGU General Assembly, Vienna, Austria, 23–28 April 2017; p. 5660.

36. Xiao, F.; Yang, Z.Q.; Lee, K.W. Photosynthetic and physiological responses to high temperature in grapevine (*Vitis vinifera* L.) leaves during the seedling stage. *J. Hortic. Sci. Biotechnol.* **2017**, *92*, 2–10. [CrossRef]

37. Valletta, A.; Salvatori, E.; Santamaria, A.R.; Nicoletti, M.; Toniolo, C.; Caboni, E.; Bernardini, A.; Pasqua, G.; Manes, F. Ecophysiological and phytochemical response to ozone of wine grape cultivars of *Vitis vinifera* L. *Nat. Prod. Res.* **2016**, *30*, 2514–2522. [CrossRef] [PubMed]

38. Mattivi, F. La stima della maturazione fenolica delle uve rosse con un nuovo metodo rapido. *ENOLOGO-MILANO-* **2007**, *43*, 89.

39. Gatto, P.; Vrhovsek, U.; Muth, J.; Segala, C.; Romualdi, C.; Fontana, P.; Pruefer, D.; Stefanini, M.; Moser, C.; Mattivi, F.; et al. Ripening and genotype controlled stilbene accumulation in healthy grapes. *J. Agric. Food Chem.* **2008**, *56*, 11773–11785. [CrossRef] [PubMed]

MDPI

St. Alban-Anlage 66

4052 Basel

Switzerland

Tel. +41 61 683 77 34

Fax +41 61 302 89 18

www.mdpi.com

Climate Editorial Office

E-mail: climate@mdpi.com

www.mdpi.com/journal/climate